The Synthetic Organic Chemist's Companion

BICENTENNIAL
1807
WILEY
2007
BICENTENNIAL

THE WILEY BICENTENNIAL—KNOWLEDGE FOR GENERATIONS

*E*ach generation has its unique needs and aspirations. When Charles Wiley first opened his small printing shop in lower Manhattan in 1807, it was a generation of boundless potential searching for an identity. And we were there, helping to define a new American literary tradition. Over half a century later, in the midst of the Second Industrial Revolution, it was a generation focused on building the future. Once again, we were there, supplying the critical scientific, technical, and engineering knowledge that helped frame the world. Throughout the 20th Century, and into the new millennium, nations began to reach out beyond their own borders and a new international community was born. Wiley was there, expanding its operations around the world to enable a global exchange of ideas, opinions, and know-how.

For 200 years, Wiley has been an integral part of each generation's journey, enabling the flow of information and understanding necessary to meet their needs and fulfill their aspirations. Today, bold new technologies are changing the way we live and learn. Wiley will be there, providing you the must-have knowledge you need to imagine new worlds, new possibilities, and new opportunities.

Generations come and go, but you can always count on Wiley to provide you the knowledge you need, when and where you need it!

WILLIAM J. PESCE
PRESIDENT AND CHIEF EXECUTIVE OFFICER

PETER BOOTH WILEY
CHAIRMAN OF THE BOARD

The Synthetic Organic Chemist's Companion

MICHAEL C. PIRRUNG

University of California, Riverside

BICENTENNIAL
1807
WILEY
2007
BICENTENNIAL

WILEY-INTERSCIENCE

A JOHN WILEY & SONS, INC., PUBLICATION

Library of Congress Cataloging-in-Publication Data

Pirrung, Michael C.
    The synthetic organic chemist's companion / by Michael C. Pirrung.
        p. cm.
    Includes bibliographical references.
    ISBN 978-0-470-10707-2 (cloth)
1. Organic compounds—Synthesis.   I. Title.
    QD262.P574 2007
    547'.2—dc22

                                            2007011296

Printed in the United States of America

10 9 8 7 6 5 4 3 2

CONTENTS

PREFACE

I hope this book will be a useful indoctrination for novice chemists making the transition from organic teaching laboratories to the synthetic chemistry research laboratory, either in academe or industry. I also attempted to assemble some of the more useful but hard to locate information that the practicing synthetic chemist needs on a day-to-day basis. My aspiration for this book is to find it (with several tabbed pages) on chemists' lab benches. Finally, I aim to remind all readers of the little details about lab work that we may learn at some point in our careers but easily forget. When you are vexed by a particularly challenging experiment, I hope that paging through this book is one approach you take to solving your problem of the day, and that it is concise enough to encourage you to do so.

I organized the book to parallel the processes involved in planning, executing, and analyzing the synthetic preparation of a target molecule. I included a new chapter not found in earlier books on this subject matter: an example of the different formats in which the synthesis of a known compound may be published. I hope this chapter assists novice chemists in translating experimental descriptions into action items for today's experiment. I also found on the Web many new and valuable electronic resources contributed by the community of synthetic chemists.

This book has been over 25 years in the making. I first learned of an effort to assist beginning experimental students in learning the ropes of research laboratory work while a postdoctoral fellow at Columbia in 1980. Clark Still was giving a mini-course to his students on how to work in the lab. This seemed a very worthwhile activity to me, knowing how inept I was in the lab at the beginning of my graduate career. That pile of handwritten notes from Still's lectures eventually grew into a typed document that was finally scanned into electronic form. Along the way, it was distributed to my graduate students and postdocs in whatever its then-current state. Lately I have searched in earnest for books with comparable content that were comprehensive and modern, and was unable to find both in one text. However, I acknowledge my debt to those who have made past attempts at this sort of synthetic chemistry boot camp. I was lucky to be able to persuade Darla Henderson at Wiley that this subject would be useful and popular, and it developed into the book presented here. I initially envisioned it would be titled "The Novice's Guide . . . ," but the opportunity she offered to echo the iconic *Chemist's Companion* penned by Gordon and Ford proved irresistible. My effort is offered in admiration of their work, and not the presumption that I can meet their high standard. I also want to be sure to recommend *The Laboratory Companion* written by Gary Coyne. It is truly a comprehensive guide to the hardware of the research laboratory, though it does not really touch on the specialized "software" of synthetic chemistry.

Finally, to the novice embarking on the study of organic synthesis, let me give you this advice: *Lasciate ogne speranza, voi ch'intrate.* This is the inscription above the Gates of Hell in Dante's *Inferno* (in the 1882 Longfellow translation, "All hope abandon, ye who enter in!"). Or, to quote a modern poet, Willie Nelson: "It's a difficult game to learn, and then it gets harder," in this case referring to golf. Synthetic organic chemistry can be one of the most frustrating, maddening, and capricious of scientific endeavors. For just this reason, success in synthesis is one of the most rewarding experiences in science. Synthesis is an intrinsically creative activity, and a chemist who does it well is often also creative in another area, be it music or cooking. If you already partake in creative hobbies, like woodworking or knitting, you can anticipate synthesis offering you similar rewards. The achievement of the total synthesis of a complex target molecule is a peak experience for synthetic chemists, often celebrated with champagne. Even the small, day-to-day successes in the synthesis lab provide a great feeling of accomplishment. Once these are experienced, I expect you will be hooked. Hopefully, this book will help your "addiction" be its most fruitful.

Michael C. Pirrung

ACKNOWLEDGMENTS

I would like to thank several reviewers and all of the graduate students and postdocs in my own lab who read and commented upon earlier versions of the manuscript. I am grateful for many figures supplied by Ace Glass. Bruce Ford provided the cover art.

CHAPTER

1

SEARCHING THE LITERATURE

It is a good idea to review some of the experimental principles and techniques that were covered in the organic teaching laboratory before undertaking the more sophisticated and less pre-defined activities of the research lab. One of the best resources for this type of information is *The Organic Chem Lab Survival Manual* (Zubrick, 2003). Awareness and observation of all of the best chemical safety practices is also essential. An excellent resource for this type of information is *Prudent Practices in the Laboratory* (Committee, 1995). Specific comments concerning chemical safety are made at some places in this book, but safety always must be foremost in the mind of the experimenter.

When aiming to obtain a particular molecule, a good appreciation of how it has been prepared in the past is essential. Electronic data retrieval tools are ideal for finding this information. The simplest way to obtain a compound is to buy it, of course.

1.1 COMMERCIAL AVAILABILITY

The SciFinder® program is a good tool for comprehensive searching of commercially available compounds. It includes the information that is published in *Chemical Abstracts*, but makes it much more accessible. This browser-like application permits searching by chemical structure, either as an exact match or based on substructure. While a comprehensive description of how to use SciFinder is not intended here, a few points should be kept in mind. Structures can be drawn in SciFinder itself, or in another chemical drawing program and pasted into the structure window in SciFinder. To avoid an unmanageable number of hits when searching by substructure, it may be necessary to limit the structure based on atoms that may be further substituted or rings that may be added. These parameters

The Synthetic Organic Chemist's Companion, by Michael C. Pirrung
Copyright © 2007 John Wiley & Sons, Inc.

must be set within SciFinder itself, with the {Lock out substitution} and {Lock out ring fusion} commands. If a compound identified in a structure search is commercially available, a small button with an orange flask will appear at the top of the compound window. A refinement of the hits from the initial structure search can also be based on their commercial availability.

For compounds of interest that show the small orange flask button, clicking it will open another window that lists the companies selling the compound and the quantities available. This last point is very important. Quite a number of compounds have been registered with *Chemical Abstracts* by small companies aiming to sell compound libraries for biological screening. These compounds, while strictly meeting the definition of commercial availability, are not in the same sense articles of commerce as are compounds sold by major reagent suppliers like Aldrich and Fluka. Information on less well known suppliers (phone, address, URL) is available by clicking a link next to each supplier listing.

Another way to search for commercially available compounds is with the ChemACX® program running on a MSWindows-based computer. A CD with a current compilation of suppliers' catalogs is needed for this application, for which periodic updates can be purchased. ChemDraw® is used to draw structures for this application. Further information on this program is available from the ChemACX manual.

It is always dangerous to refer to a Web-based resource in a book that should have long-term value, since Web pages seem to come and go at a rapid pace. However, some equivalent resources may arise to take their place. The case at hand involves the Web site of the Sigma-Aldrich Company, which is an excellent free resource to use in locating commercially available compounds. From the main page, a few clicks brings the browser to a search page. Searching by structure is among the options there. Chemical structure drawing on Web pages works fine, provided that the plug-ins are installed in your browser, but can be clunky. Better in this case is to enter the structure as a SMILES string. Many approaches have been taken to manipulate chemical structures electronically, with one basic machine-readable nomenclature being the SMILES format. SMILES stands for Simplified Molecular Input Line Entry Specification. The SMILES description of the structure sought can be obtained from a drawn structure in a program like ChemDraw using the command {Copy as SMILES}. Upon pasting the SMILES string into the Sigma-Aldrich search page and loading it, the structure appears. It can be searched for using several options available on the Web site. A nice aspect of this search is that it encompasses all of the Aldrich sister companies, including Sigma and Fluka. Aldrich is a company that has always provided excellent service to chemistry, and many chemists keep an Aldrich catalog at their desks because of the wealth of data it includes on all of the compounds listed.

1.2 LITERATURE PREPARATIONS

If a compound cannot be purchased, there may be a known preparation of it. It is almost always preferable to at least begin with a method that others have used to prepare a compound, rather than trying to invent one from scratch. Identifying the specific publications in which a compound has been *prepared* among many more papers in which it has been mentioned in some way is straightforward when using SciFinder. The {Refine} command allows those sources to be selected. If a particular method used to prepare a compound is sought, it may prove useful to search by reaction rather than by compound. This is done simply—multiple structures are entered into the chemical structure search window, and a reaction arrow is drawn from the starting material(s) to the product. SciFinder will then label the molecules that it understands to be the reactant(s) and product(s). Searching for this reaction can next be initiated. The number of hits is often quite large, but adding qualifiers can reduce them to a manageable number. For example, reactions can be selected by applying the {Refine} command based on yield, or the number of steps, or the general type of chemical transformation. Another option that often reduces the possibilities is to identify which atoms of the reactant correspond to the same atoms in the product. This may seem obvious to chemists, but it is not obvious to this software.

A preparation of a compound that was conducted in one's own laboratory is most often preferred as a starting point for today's preparation. In some cases the chemist who actually did the prep may be available for consultation, or at least his or her research notebook pages can be consulted. A properly kept research notebook (see Chapter 7) is almost always more useful than any literature preparation because it gives details that never appear in a publication, such as actual chromatograms and spectra, drawings or pictures of apparatus, and properties of chromatographic or distillation fraction.

The best literature methods will be found in the compilation *Organic Syntheses*. These procedures are generally aimed at synthesizing compounds on a fairly large scale (grams or larger, rather than milligrams). They are distinguished from essentially all other literature preparations by having been checked in the laboratory of another experienced synthetic chemistry group. The little details that can be important to success can be incorporated into the published procedure through the experience of the "checkers" in doing the experiment targeted for today, reproduce someone else's preparation. The only drawback of *Org. Syn.* preps is that so few of the known compounds one might need have been through its rigorous review process. Even if the exact target compound is not in *Org. Syn.*, a prep of a related compound might provide a good starting point if today's target is not too different structurally. Structure, keyword, title, author, registry number, molecular formula, and chemical name searching of the compilation are available on the Web site for *Org. Syn.* at http://www.orgsyn.org/.

Another source that is focused on preparative chemical procedures is the Synthetic Pages Web site. It also provides information not usually found in journal articles, such as troubleshooting tips, frequently encountered problems, and known variations with scale. It offers the virtue that it is interactive, with continual updates and comments from users. Its main limitation is that as of October 2006 there were just over 220 procedures in the database. This free database is found at http://www.syntheticpages.org.

Next preferred for the preparation of a desired compound is one that has been described in a "full paper" (one that gives experimental details, as in *J. Org. Chem.*). These experimental details are increasingly being relegated to the electronic/Web version of the journal, often called Supporting Material or Supplementary Information. This material is often available to all, even those without electronic access to the main journal. The techniques for searching for this literature include *Chemical Abstracts* as described above, as well as specialized sites maintained by each publisher for their own journals.

A particularly useful Web site to locate information concerning specific reactions is the *Encyclopedia of Reagents for Organic Synthesis*, in its online form, e-EROS. It allows searching of a database of more than 50,000 reactions and more than 3000 of the most commonly used reagents by chemical structure and substructure, reagents, conditions, and reaction type. The URL is http://www3.interscience.wiley.com/cgi-bin/mrwhome/104554785/.

Least informative concerning the preparation of a target compound are the so-called communication or letter publications, which do not include experimental details. In some cases experimental descriptions may be provided in Supplementary Information. If not, the author might provide some detail about reaction conditions in the written narrative or on the reaction schemes. It is always worth contacting the senior author to find out if experimental details might be obtained. It is often easy to find and contact people electronically. Still, the novice experimenter may have little to go on, and reproducing results from these types of papers is widely regarded as difficult at best.

Finally, it should be recalled that the electronic versions of journals are in some cases only available for relatively recent editions. Electronic coverage of compounds and reactions earlier than that time cannot be assured. Thus some traditional book-based literature search methods may be required if the reaction of interest has a long history.

CHAPTER

2

REAGENTS

It is wise to check the purity of all reactants before starting a reaction, especially when trying new reactions. Otherwise, if a reaction fails, it is difficult to know if the cause is the reagents or a fault in the procedure. Analytical methods could include titration, NMR spectroscopy, thin-layer chromatography, or gas chromatography, depending on the types of contaminants that might be involved. These baseline data on reactants will also be useful in analyzing reaction mixtures so that starting materials can be easily identified. Some reactive reagents (trimethylsilyl triflate, acid chlorides) may not be easily analyzed. One might be surprised at the effectiveness of NMR spectroscopy for determination of reagent quality. For example, the chemical shifts of the methyl groups of acetic acid and peracetic acid appear at δ 2.16 and 2.22, so an NMR spectrum of a commercially available peracetic acid reagent solution readily provides its titer.

A classic text on the purification of reagents is known in many labs simply as Perrin (Armarego and Perrin, 2000). A major part of this text is an amazing compound-by-compound listing of recommended methods for purification. Another resource is *Practical Organic Chemistry* sometimes called simply Vogel (Furniss et al., 1989). Although this text is somewhat dated in terms of techniques and equipment, best practices for reagent purification change slowly. If a reagent should be purified, if it is not too reactive, and if it is somewhat volatile, distillation is a frequent choice.

2.1 SHORT PATH DISTILLATION

This method entails a simple distillation that is useful for quantities of material to be purified in the gram to several grams range. The short path distillation apparatus is a simple, integrated still head that

The Synthetic Organic Chemist's Companion, by Michael C. Pirrung
Copyright © 2007 John Wiley & Sons, Inc.

accepts a mercury immersion thermometer and receiver (Fig. 2.1) and can be placed under vacuum or an inert atmosphere. The pot is heated with an oil bath, and the apparatus is wrapped in aluminum foil to minimize radiative heat loss and the overheating necessary to get material to distill over. A crude fractionation of the distilland can be done, provided the boiling points are well separated. To distill under reduced pressure, a multiple receiver adapter (Fig. 2.2; otherwise, known as a "cow," for obvious reasons) can be used that permits the distillate to be directed, merely by rotating the cow, to different receivers. This apparatus can be adapted for fractional distillation of larger quantities of reagent by adding a column such as the Vigreux

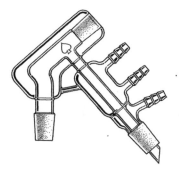

Figure 2.1 A short-path still head.

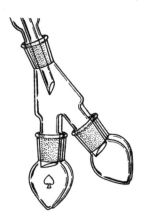

Figure 2.2 A cow adapter for vacuum distillation with collection of several fractions.

type (Fig. 2.3). The column must be thermally insulated. A typical height equivalent to a theoretical plate (HETP) value for a Vigreux column is 10 cm. A 50 cm column is therefore capable of providing about 5 theoretical plates, which can acceptably separate (ca. 95% pure) compounds with a boiling point difference of 30°C.

When performing any distillation at reduced pressure, it is important to have an estimate of the boiling point of the desired material, which of course is pressure dependent. It is possible to translate a known boiling point at one pressure to the boiling point at another pressure by using a vapor pressure nomograph (Fig. 2.4).

To use this nomograph, given the boiling point at atmospheric pressure, place a straightedge on the temperature in the central column. Rotating the straightedge about this temperature will afford the expected boiling point for any number of external pressures. Simply read the temperature and the corresponding pressure from where the straightedge intersects the first and third columns. For example, choose a boiling point at atmospheric pressure of 400°C. Using the nomograph and this temperature for reference, rotating

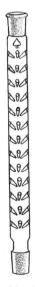

Figure 2.3 A Vigreux column for fractional distillation with moderate separating power.

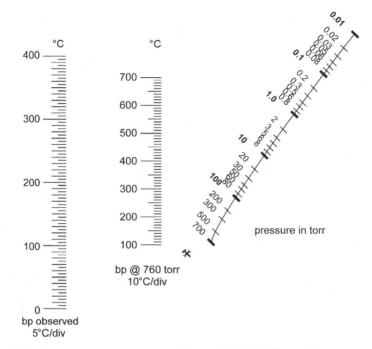

Figure 2.4 A vapor pressure nomograph permits the boiling point at any pressure to be predicted based on a known boiling point at a known pressure.

the straightedge about this temperature will afford a continuous range of expected boiling points and the required external pressures necessary to achieve the desired boiling point. At a pressure of 6 torr, the expected boiling point will be 200°C. Likewise, our compound boiling at 400°C at 1 atm would be expected to boil at 145°C at 0.1 torr external pressure.

Electronic versions of the vapor pressure nomograph are available on the Web, two nice ones being:

http://www.sigmaaldrich.com/Area_of_Interest/Research_
    Essentials/Solvents/Key_Resources/nomograph.html.

http://www.chemsoc.org/exemplarchem/entries/pkirby/
    exemchem/Nomograph/Nomograph.html.

2.2 AMPULES

Some reagents (e.g., methyl triflate, ethylene oxide) are supplied in sealed glass ampules (Fig. 2.5). Two possible reasons for this packaging are the compound is very reactive with the atmosphere, or the compound is quite volatile and conventional closures are inadequate to contain it. It is usually not a good idea to use only a portion of the material in the ampule and/or attempt to reseal it. The best strategy is to set up the reaction on a scale such that all of the reagent in the ampule is consumed in today's reaction. Second best would be to transfer the unused reagent into a sealable storage bottle or flask under an inert atmosphere. The neck must be broken off the ampule to access the reagent. Usually there is a marked line on which the break is intended to occur, but it is always good practice to score the glass with a glass file or diamond glass cutter to be certain the neck breaks cleanly and easily. When the reagent is reactive with the atmosphere, this should be done with close access to an inert atmosphere or actually under an inert atmosphere (e.g., inside a glove box). When the reagent is volatile, the bottom of the

Figure 2.5 Reagents packaged in glass ampules. Wheaton Science Products, a division of Wheaton Industries, Inc.

ampule should be cooled before opening to minimize the internal pressure.

2.3 REAGENT SOLUTIONS

A wide variety of reagents are available from commercial suppliers as solutions. This makes for easy measurement of even quite air sensitive reagents by volume using a syringe (see Section 8.4). Reagent solutions are particularly useful for organometallic reagents that are better prepared on a large scale, but many, many air-sensitive and sophisticated reagents are today available in this way. One factor against using these reagent solutions is their high expense on a molar basis (compared to preparing them in the lab), and another is the reliability of their titer. Even a new bottle from a supplier may have significantly different reagent concentration than the label claims. When a chemist accepts the titer of a purchased reagent solution, he or she is relying on someone else to have performed their experimental work correctly. The concentration listed is usually a minimum, but regardless, it is always a good idea to check the concentration when possible. Of course, over time, with storage and with use, titer may decrease, so periodic titration may be necessary.

Bottles containing these compound solutions frequently come with closures intended to preserve an inert atmosphere, such as the Aldrich Sure/Seal.™ These involve a septum covered with a fiber mat fixed under a metal cap similar to that on a soda bottle (the crown cap). A conventional plastic screw cap goes on the bottle over this closure. The Sure/Seal works fairly well, but many chemists add their own precautions. The sleeve of a regular rubber septum (Section 4.5) can be put on (upside down) over the Sure/Seal and wired on; a smaller second septum can be inserted inside the first and its sleeve folded over. Stretching of Parafilm M® over this septum is also a good idea. Storage in a refrigerator helps to maintain concentrations of solutions in more volatile (i.e., pentane, ether) solvents. MeLi is not stored in the refrigerator, as it precipitates out of solution and this changes the concentration. Most organometallics stay as clear solutions as long as they are good, with white precipitates (metal hydroxides) indicating some loss of reagent. All bottles of reagent solutions should be maintained with an inert atmosphere in the headspace (the region above the solution).

2.4 TITRATION

Organometallic reagents that are also strong, stoichiometric bases like Grignard and lithium reagents may be titrated in at least three ways: against aqueous acid, against diphenylacetic acid, and against alcohol. The first is performed by quenching an aliquot of the solution into water and titrating the hydroxide produced with 1 M acid, using phenolphthalein or bromocresol purple as an indicator. This method has the disadvantage that it does not discriminate between the organometallic and hydroxide ion that may be present in the solution from its reaction with moisture (eq. 1). This hydroxide may be subtracted by using the Gilman double titration. Hydroxide can be measured independent from the organometallic because the latter is destroyed without the production of base by addition of dibromoethane (eq. 2). The amount of hydroxide remaining is determined by titration as above. Another disadvantage of this method is the two-phase system that is produced when mixing organic and aqueous solvents. This makes the endpoint more difficult to determine than in a standard aqueous titration.

$$MeMgBr + H_2O \longrightarrow CH_4 + HOMgBr \tag{1}$$

$$MeMgBr + (BrCH_2)_2 \longrightarrow MeBr + C_2H_4 + MgBr_2 \tag{2}$$

The second method is performed as follows: recrystallized, oven-dried diphenylacetic acid is weighed and dissolved in anhydrous tetrahydrofuran in a round-bottom flask containing a stir bar and kept under nitrogen. The amount of diphenylacetic acid used is determined such that less 1 mL of the organometallic solution (based on its estimated concentration) will be required, permitting a 1 mL syringe graduated in hundredths to be used for the titration. The syringe is filled (Section 8.5) with the organometallic solution and it is slowly dropped into the flask. Once the diphenylacetic acid is fully converted to the carboxylate, further addition of organometallic produces the yellow dianion (eq. 3), which indicates the endpoint. Because diphenylacetic acid can be neutralized by hydroxide in the organometallic this process can lead to some error.

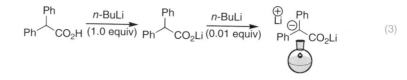

(3)

The third method uses an alcohol such as *sec*-butanol as the acid, and since it will protonate only the organometallic, no correction for hydroxide is necessary. The other key to this method is the indicator, phenanthroline or bipyridyl. These have the property that they form colored complexes with lithium and magnesium reagents. In their presence, a brown color is usually observed. In their absence, a yellow or clear solution is seen. The titration is performed with an automatic burette (Fig. 2.6) under an inert atmosphere. The automatic burette has a lower reservoir that can be filled with titrant (ca. 0.6 M *sec*-butanol in anhydrous xylene, determined volumetrically). The section of the burette that we would recognize as a burette is filled with titrant by pumping a rubber bulb and forcing liquid up into the burette with pressure. It is a convenient apparatus to store and maintain the titrant solution and encourages frequent titration.

Anhydrous tetrahydrofuran (ca. 3 mL) and the indicator (20 mg) are added to a round-bottom flask with a stir bar. A known amount of organometallic is added, and the color should develop. Titrant is added from the automatic burette until the color is just discharged. Provided that the endpoint is not greatly exceeded, replicate titrations can be performed in the same flask to obtain a more precise value.

It is also certainly possible to prepare solutions of reagents oneself. A particular instance where this may be useful and for which commercial sources are often not available is volatile compounds. A classic example is anhydrous HCl in methanol. This solution cannot be purchased (though other solutions of HCl in organic solvents can be), and one would not want to store it for long periods because it is so corrosive, but it is perfectly reasonable to prepare such a solution and keep it in the hood for several days. It is prepared by carefully bubbling gaseous HCl from a tank (see tank methods in Chapter 3) into a known volume of dry methanol in an ice-cooled, tared flask. (CAUTION: HCl has a substantial heat of solution, so this process is *very* exothermic. It is important to very carefully lower the sparging

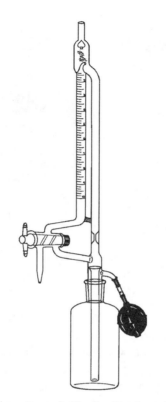

Figure 2.6 An automatic burette useful for the titration of organometallic reagents. The reservoir contains a nonhygroscopic solution of *sec*-butanol in xylene at a known concentration. The bulb is squeezed to fill the burette, and the tip is inserted through a serum cap into a flask under inert atmosphere and containing an ethereal solvent in which the titration is conducted.

tube into the solution to avoid drawing methanol back into the gas line.) After the reaction has returned to ambient temperature, the flask is weighed and the HCl concentration is determined. Another example is volatile organic compounds like methoxyacetylene or ethylene oxide that have boiling points lower than about 30°C. It is very difficult to measure them neat in a syringe or on a balance, even when chilled, because they evaporate at such a rapid rate. Dissolving them in a reaction solvent and determining the titer, either by a direct method (e.g., integration in NMR spectroscopy) or by weight as was done for HCl/MeOH, can make their use much more convenient.

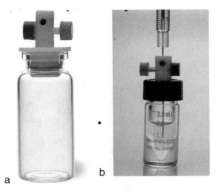

Figure 2.7 (*a*) A Mininert valve inserted into a vial. The small dot is a cylinder of rubber septum material through which a needle is inserted. (*b*) When the green button on the Mininert valve is pushed in, the pathway is clear for a needle to be inserted into the liquid to withdraw reagent. When the needle has been removed, the red button on the valve is pushed in to seal the container from the external atmosphere.

Sometimes repeated access (while maintaining an inert atmosphere) to homemade reagents like Grignards that are not commercially available is required. One way to achieve this is with Mininert valves (Fig. 2.7) that are available in sizes that fit small (100 mL) reagent bottles, 5 to 30 mL reaction vials, and 14/20 joints. Another option is a three-way stopcock fitted with a joint and attached to a flask. To remove the reagent, a nitrogen line is attached to the sidearm of the stopcock. The stopcock is opened, a needle is inserted through it, and the liquid is withdrawn (see technique below). The nitrogen flow maintains a blanket of inert atmosphere around the stopcock and replaces the volume of the withdrawn liquid.

Any time a reagent in any form is bottled, an accurate label should be created that provides as many details as possible concerning its date of bottling, contents, and concentration. This information is most secure and robust when written *in pencil*, since inks readily run when contacted by the pervasive organic solvents.

2.5 REAGENT STORAGE

For reagents that are obtained in standard screw-cap bottles, the cap should be wrapped in Parafilm M following use. Parafilm is a thin, elastic sheet with low water permeability that is resistant to hydrochloric, sulphuric, and nitric acids, sodium hydroxide, potassium permanganate, and ammonia solutions, and ethyl alcohol, isopropyl alcohol, and acetonitrile. It is not stable to acetone or chlorinated hydrocarbons. The purpose of the Parafilm wrap is as much to keep the reagent in during storage as it is to keep the environment out. If a reagent is light-sensitive, the bottle should be wrapped in aluminum foil; brown glass bottles are rarely effective in fully preventing light penetration. If a reagent has a stench, the bottle should be placed inside a plastic bag.

Storage conditions depend on the particular reagent. Reagents that are not volatile, have low or no moisture or oxygen sensitivity, and are thermally stable can be stored on a shelf, ideally in a cabinet vented to the hood system. Reagents with moisture sensitivity but no oxygen sensitivity can be stored in a desiccator. Reagents that degrade when exposed to moisture or oxygen in the atmosphere should be stored under an inert gas. Reagents that are volatile or are otherwise heat-sensitive should be stored in the refrigerator. Allow reagents removed from a refrigerator to warm to room temperature before opening the bottle, to avoid drawing air into the bottle. Few reagents truly require freezing, unless so labeled. Never place aqueous solutions in the freezer.

2.6 SUBTLE REAGENT VARIATIONS

Some aspects of reagents can have a profound effect on their reactions, and in some cases these traits are uncontrollable. For example, many common organometallic reagents like alkyl lithiums are purchased for convenience and because they can be hard to make. Does it matter whether the alkyl lithium was made by the commercial supplier from the bromide, the chloride, or the iodide? Does it matter whether the alkyl lithium is supplied in ether or tetrahydrofuran? Often, it does matter. Why should this be so? In some cases, it has been shown that dissolved metal salts (LiCl, LiBr, and LiI) affect the outcome of a reaction. These salts have different solubilities in ether and tetrahydrofuran (LiI is the most soluble), so the solvent (which will certainly be known) and alkyl halide precursor (which may not be known) of the purchased alkyl lithium can strongly affect the salts present in the solution and therefore the outcome of the reaction. Wittig reactions can experience a similar effect, where the stereochemistry is dependent on the counter ions of both the base used to generate the ylide and the phosphonium salt. Using sodium and potassium bases leads to salts that are insoluble and cannot affect the reaction. Such "salt-free" Wittig reaction conditions are commonly among the most stereoselective.

2.7   DANGEROUS REAGENTS

There are certain classes of compounds that are intrinsically unstable and/or prone to explosion. It is crucial to be aware of these classes, since examples might be prepared in the course of research that had never before been known. The chemist will not have a material safety data sheet as a warning, but the explosion will be just as dangerous. The classes include acetylenes, acetylide salts, and polyacetylenes; hydrazoic acid and all azides, organic or inorganic (excepting sodium azide, which is safe); diazonium salts and diazo compounds; organic or inorganic perchlorates; nitrate esters of polyols; metal salts of nitrophenols; nitrogen trihalides; and peroxides.

2.8   REAGENT PROPERTIES

It is always useful to be aware of some of the physical properties and constants for the reagents used in a reaction. For example, an accurate value for the density can be very helpful in measuring reagents by syringe. Knowledge of the freezing points of all solvents and reagent is needed to avoid freezing them during low-temperature reactions. This type of information is frequently available for more common compounds. One obvious source is a handbook, like the *CRC Handbook of Chemistry and Physics*, but another common resource is chemical catalogs. These catalogs have incorporated more and more compound-specific information over time, and many chemists keep a chemical catalog or two at their desks. Aldrich even calls its catalog the *Aldrich Handbook*.

CHAPTER

3

GASES

A variety of compressed gases are available. They are stored in containers ranging in size from lecture bottles (Fig. 3.1; about the size of a 250 mL graduated cylinder) to small and large tanks or cylinders (Fig. 3.2). Each type has a specific method for delivering the gas.

3.1 LECTURE BOTTLES

Lecture bottles usually have one female threaded fitting into which a needle valve (Fig. 3.3) is attached. With the needle valve closed (fully clockwise), a nut below the fitting is screwed in (down, clockwise), which opens a pressure valve. Flow is then controlled with the needle valve.

3.2 TANKS OR CYLINDERS

Two types of gas delivery are available, with unique fittings for each gas. This minimizes the possibility that an incorrect gas might be introduced into a system intended for another gas.

A needle valve screws onto the tank valve and has a hose adapter for output. The needle valve is closed (fully clockwise) and a hose is attached. The tank valve is opened by turning the square nut on the top of the tank counterclockwise with a crescent wrench. Delivery is controlled by the needle valve. Examples: NH_3 and HCl.

Regulators (Fig. 3.4) are used for providing gases at constant pressure. A regulator has two gauges and a wing screw in front and screws onto the tank valve. It sometimes has a needle valve at the output, but if it does not, consider adding one. After attachment to the cylinder valve, the wing screw is turned to the fully closed,

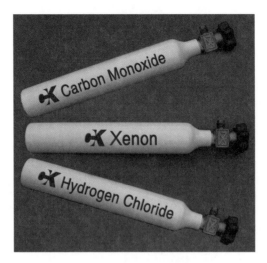

Figure 3.1 Lecture bottles used to hold and deliver small quantities of reagent gases.

Figure 3.2 Large gas cylinders typically used for gases like He, N_2, and H_2. Photo provided by Airgas, Inc.

Figure 3.3 A brass valve for a lecture bottle.

Figure 3.4 A gas pressure regulator for a large cylinder. Photo provided by Airgas, Inc.

counterclockwise position. The tank valve is opened, causing the right-hand valve to show the pressure of gas contained in the tank. With the needle valve closed, screwing the wing screw in (clockwise) depresses a diaphragm that opens to release the gas. The left-hand gauge shows the delivery pressure. Opening the needle valve delivers the gas. Examples: H_2, He, and air for GC; N_2 and Ar for inert gases and HPLC degassing.

3.3   GAS SAFETY

A few general safety precautions should be observed when working with gas tanks. Always transport tanks with a cart. Tanks should be chained or strapped to a wall or table at two points at all other times. Any tank not in active use should bear its protective, screw-on metal cap. Toxic and corrosive gases (e.g., NH_3, HCl) are best obtained in small tanks that can be placed in hoods. Any gas at high pressure carries the risk of asphyxiation, since if all of the gas in a tank were to be released in a closed area, its volume would be large enough to displace all the air.

CHAPTER

4

REACTIONS ON A SMALL SCALE—1 TO 25 MMOL

Given the title of this chapter, something should be said about how the chemist selects the scale on which a reaction is conducted. One principle is that a reaction should be conducted on as small a scale as possible. This is so that, should the reaction fail, the loss of starting materials could not be too costly. "Cost" in this context can be literally the dollars spent for the purchase of starting materials or, more seriously, the chemist's time and effort in making the starting materials if they are the product of other synthetic reactions. For the latter reason chemists who are executing lengthy syntheses become highly adept at conducting reactions on small scales (a milligram or less!) "at the frontier." Of course, once one has success with a particular reaction, the issue of scaling up to bring more material through the synthesis arises. An oft-mentioned rule of thumb is that a reaction should never be scaled up more than 4× because it may not work as well at the larger scale, and further modifications of the procedure may be needed. This rule is likely honored as often in the breach as the observance, since it is obviously difficult to accumulate a significant quantity of a target compound if scaling up starts at the 1 mg level and is limited to fourfold increases.

Why should the outcome of a reaction be dependent on the scale at which it is conducted? There are several reasons. Mixing and heat transfer may be less efficient when reaction volumes are larger. Heterogeneous reactions that depend on reagent transfer between two phases are notorious for problems in scale-up. While a chemical step might have been limiting at the smaller scale, the efficiency of mixing might become limiting at the larger scale. The apparatus used for a reaction can never be identical at different scales—think of scaling up a 10 mL syringe by 100×. For the same reason reagent concentrations may be increased upon scale-up to obtain greater productivity from a reaction, which may affect the yield.

The Synthetic Organic Chemist's Companion, by Michael C. Pirrung
Copyright © 2007 John Wiley & Sons, Inc.

When working from a well-described literature procedure, it may be desirable (subject to the foregoing caveats) to try to follow its scale closely. Some experimental procedures are expected to be more efficient on a larger scale, especially those including processes in which material transfer losses can occur, like distillation.

4.1   REACTION FLASKS

Many reactions are run under an inert atmosphere of nitrogen or argon, are operated at other than room temperature, and require addition of reagents without exposing the reaction to the atmosphere. These three functions are well served by a three-neck flask (Fig. 4.1). The flask may be connected to an inert atmosphere manifold (see Section 4.4) through a gas inlet, fitted with a rubber septum for addition of reagents by syringe, and fitted with a thermometer that extends into the reaction solvent. This latter point is essential not only to observe the *exact* temperature of the actual reaction solution (rather than just the bath that surrounds the flask), but also

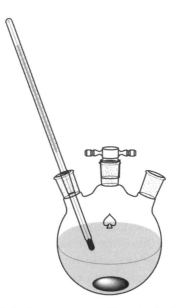

Figure 4.1 A glassware setup for a typical reaction.

to observe changes in reaction temperature as it proceeds. That is, upon the addition of a reagent, an exothermic reaction will show a temperature increase that can be observed if there is a thermometer in the solution. Because of the large thermal mass of a temperature-controlling bath, such exotherms are not easily observed via the bath temperature.

It is important to consider final reaction concentration, scale, and therefore volume in selecting flask size. Reasonable concentrations for fast bimolecular reaction rates might be in the 0.1 to 1 M range, so if a 10 mmol reaction is being set up, the final volume will be from 10 to 100 mL. On the other hand, reactions in which processes with unimolecular kinetics are to be favored over those with bimolecular kinetics (e.g., an intramolecular vs. an intermolecular reaction) might be conducted at high dilution, meaning final concentrations closer to 0.001 M. In order to prevent sloshing of contents during mixing and to potentially allow the workup to be conducted in the reaction flask itself, the flask should be at least double the volume of the reaction, and ideally treble its volume.

Almost all reactions will be conducted in a glass apparatus. The most often used glass is Pyrex®, a borosilicate glass that is one of the most inert substances to chemical reagents. The only exceptions to this generalization are hydrofluoric acid (HF), hot phosphoric acid, and hot alkali. Of these, hydrofluoric acid is the most serious problem. Attack on the glass can occur even when a solution contains only a few parts per million of HF. Other fluoride reagents can attack glass but usually to a lesser extent than HF. These reagents dissolve glass because of the very strong Si–F bonds that are formed. Phosphoric acid and concentrated alkali solutions are not a problem when cold; they only corrode glass at elevated temperatures. Alkali solutions up to 30% by weight can be handled safely at ambient temperature. For reactions that use reagents that attack glass, apparatus made of Teflon® or other plastics can be used. Borosilicate glass can be used safely at temperatures up to 232°C provided that it is not subjected to rapid changes in temperature (>120°C). The strength of borosilicate glass increases with decreasing temperature, and it can be safely used at cryogenic temperatures.

While the foregoing considerations address whether the glass can tolerate the reaction, the complementary consideration is whether the reaction can tolerate the glass. A number of reactions are dependent on the exact properties of the flask surface. The reasons for these observations are rarely established, but they are usually thought to be due to trace levels of acid on the glass from previous reactions or cleaning protocols. The chemical structure of glass includes surface Si–OH groups, but the acidity of Si–OH groups (in silica) is fairly weak (pK_a of about 7, over 100-fold less acidic than acetic acid). It is unlikely the glass itself can provide acidity that affects many chemical reactions.

Two common methods are used to try to prevent interference with a reaction by the glass: base washing and silanization. Base washing is quite simple, typically 28% aqueous ammonia solution is used to wash out the flask just before drying it directly in the oven (i.e., without a rinse). Silanization converts the Si–OH groups to Si–O–SiR groups. The reagents used for glass silanization include dichlorodimethylsilane (DCDMS), trimethyldimethylaminosilane, and chlorotrimethylsilane (TMSCl). The former gives a more robust protection of the silicon oxide surface because it is attached by two Si–O–Si bonds.

Silanization is performed with a 5% (v/v) dichlorodimethylsilane solution in toluene or dichloromethane made up on a daily basis. Glassware at room temperature can be soaked in this solution, or it can be filled with this solution. After a 5 minute to 2 hour treatment, the glassware is removed from the solution, or vice versa. The glass is immediately rinsed with pure solvent. Exposure to moisture in the air is minimized by proceeding immediately to the next step. The glassware is covered or filled with methanol for 15 to 30 minutes. The methanol is drained off and the glass is blown dry using nitrogen.

4.2 STIRRING

Certainly the most common (and easiest) stirring method is with a magnetic stirring bar and motor or stirring plate. This method works well for reactions in free-flowing solutions. Magnetic stirring bars come in a wide range of sizes, shapes, and coatings. Cylindrical stir bars with a fulcrum in the middle are excellent for Erlenmeyer flasks, but they do not spin well in round-bottomed flasks. For these containers, football-shaped stir bars (available for flasks 25 mL and larger) often perform better. Shorter cylindrical stir bars sometimes fit in a round-bottomed flask and spin well. For 10 mL or smaller flasks, vials or test tubes, tiny cylindrical bars stir (often called "fleas") work well. The most common stir bars are Teflon coated (Fig. 4.2). Teflon is a very tough substance, and these stirring bars will stand up to almost any reaction conditions, except very strongly reducing conditions (i.e., alkali metals in ammonia, $LiAlH_4$). These reagents will rapidly turn Teflon-coated stir bars black. They still seem to work all right, so workers who do these types of reactions often keep their black stir bars just for this type of reaction. Alternatively, glass encapsulated stir bars (Fig. 4.3) can be used.

Magnetic stirring may not work well, typically where the forces of friction, inertia, and solution viscosity are greater than the force exerted on the stir bar by the magnet in the magnetic stirrer. When this occurs, strategies to solve the problem include moving the stirrer

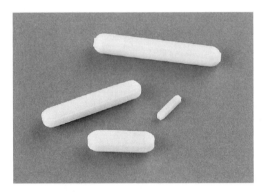

Figure 4.2 Teflon-coated magnetic stir bars.

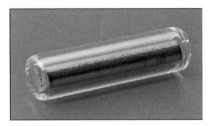

Figure 4.3 A glass-encapsulated magnetic stir bar.

closer to the bottom of the flask, using a stirrer with a stronger magnet, or using a larger stir bar. The magnetic stirrers with strong and weak magnets are often well known to all the chemists in the lab. Magnetic stirring may fail completely for reaction mixtures that are viscous, in which precipitation occurs, or that have larger reaction volumes. In such cases mechanical stirring may be required. A glass rod with paddles on its end is inserted through a bearing (oiled with mineral oil) in a ground glass joint (Fig. 4.4). A variable-speed electric motor clamped above the flask is attached to the glass rod via a short piece of rubber tubing (slipped over it) or through a flexible shaft. This provides some flexibility in the connection so that the vibrations and movement that are natural to the operation of the motor do not place stress on any of the glass in the apparatus. The speed is controlled by the motor's integral control or a Variac or

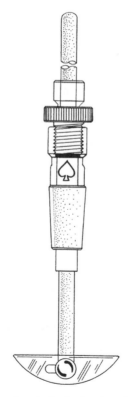

Figure 4.4 An overhead mechanical stirring bearing for a ground-glass joint.

Figure 4.5 A Variac or Powerstat sets the output voltage to AC heating devices to control their temperature and motors to control their speed.

Powerstat (Fig. 4.5). A Variac is a variable AC voltage transformer (like a heavy-duty laboratory light dimmer).

Finally, it should be kept in mind that stirring affects only the rate of mixing of liquids and cannot affect the intrinsic rate of the reaction. Therefore, for reactions in a homogeneous solution, stirring may be unnecessary. It is rare that an experimental description omits stirring, however.

4.3 GLASS JOINTS

Several considerations apply to use of ground glass joints. In the past it was essential to use a good stopcock grease on every joint in order to ensure that the apparatus could be disassembled following the reaction, especially one involving a temperature change. These greases are typically silicon polymers that are quite soluble in hydrocarbon and halocarbon solvents, so it is easy to contaminate a reaction product with stopcock grease. This may be indicated by a yield greater than 100%, a high R_f spot on TLC, or signals in crude NMR spectra (like tetramethylsilane, the Me–Si signals from the grease are around 0 ppm (δ) in the proton NMR spectrum). An effective (though expensive) alternative is Teflon sleeves (specific to each joint size) or Teflon tape. Some glassware may be supplied with small glass "dogs" or hooks adjacent to each joint that enable the apparatus to be firmly held together with rubber bands, springs (sometimes supplied), or even copper wire. Alternatively, clips are available

Figure 4.6 Metal clip for a ground-glass joint. Photo courtesy of Kimble/Kontes.

(Fig. 4.6) to hold the joint together, though these are likely less reliable and robust than the dogs and can sometimes interfere with other glassware and joints on the apparatus.

4.4   INERT ATMOSPHERE

Reactions are potentially susceptible to three minor components of air: oxygen, water, and carbon dioxide. Organometallic reagents are particularly susceptible to reaction with oxygen (eq. 4) and water (eq. 5). Any reaction involving a basic reagent more basic than hydroxide ion may be impeded by the presence of water (eq. 6). Thus knowing the relative pK_as of a variety of organic functional groups and reagents is important. The acidities of a range of compounds in both aqueous and nonaqueous media are provided in Appendixes 6 and 7. Carbon dioxide reacts with hydroxide (eq. 7), alkoxides (eq. 8), and amines (eq. 9). Because they are gases, keeping oxygen and carbon dioxide away from a reaction is mainly a matter of using an inert atmosphere, as described below. Water may be present in microscopic amounts even on apparently dry glassware and other apparatus. Water is typically removed in one of two ways. Equipment can be dried in a drying oven (>130°C) overnight and then cooled in a desiccator. Some workers take glassware directly from the oven to the reaction bench (using insulated gloves or tongs) and cool it under a flow of inert gas. For equipment that was not dried in a drying oven, it may be dried with a flame (or, in labs that discourage open flames, by using a heat gun). Typically this method is used for a completely assembled apparatus, and offers the virtue that its components that are not usually oven-dried, like Teflon tape and serum stoppers, become dried. To be absolutely certain that there is no significant water in an apparatus, it can be cooled below −40°C. Any trace of condensation signifies that water is present and the drying process must be repeated.

$$PhLi \quad + \quad O_2 \quad \longrightarrow \quad Ph\diagdown O\diagdown O\diagup Li \qquad (4)$$

$$PhLi \quad + \quad H_2O \quad \longrightarrow \quad PhH \quad + \quad LiOH \qquad (5)$$

$$t\text{-BuOK} \quad + \quad H_2O \quad \longrightarrow \quad t\text{-BuOH} \quad + \quad KOH \qquad (6)$$

$$KOH \quad + \quad CO_2 \quad \longrightarrow \quad KHCO_3 \qquad (7)$$

$$KOMe \quad + \quad CO_2 \quad \longrightarrow \quad MeOCO_2K \qquad (8)$$

$$2\ MeNH_2 \quad + \quad CO_2 \quad \longrightarrow \quad MeHN\text{-}CO_2^{\ominus}\ ^{\oplus}H_3N\text{-}Me \qquad (9)$$

The selection of nitrogen or argon as the inert atmosphere is based on a few considerations. Argon is about 3× more expensive than nitrogen, and argon can only be obtained from a tank, whereas nitrogen can come from a tank or "house" nitrogen, usually obtained by the evaporation ("boil-off") from a liquid nitrogen dewar. Argon is denser than air, which is advantageous in that argon sinks to the bottom of any container and displaces the air. Argon is also not reactive with lithium metal, whereas nitrogen does undergo a slow reaction with Li°. Nitrogen is usually used without problem for Li°/ NH$_3$ reductions, but the formation of organolithiums from alkyl halides and Li° is properly done under argon.

Manifolds are very helpful to running reactions under inert atmosphere and can be simple to set up. They require a source of pure gas (nitrogen or argon). A simple inert gas manifold is assembled starting with a gas drying tower (Fig. 4.7) filled with DRIERITE, a calcium sulfate desiccant that can be obtained with a color self-indicator, or KOH (*never* with bubbling through H$_2$SO$_4$). It is connected via tubing to a series of three-way stopcocks and terminates with an oil bubbler (Fig. 4.8). Tubing emanates from each stopcock, terminating with syringe needles or gas inlets (Fig. 4.9). The bubbler permits the chemist to see that there is positive pressure and a moderate flow of gas in the system and permits volume/pressure increases to be quickly equilibrated. A bubbler provides a means to have a system isolated from the atmosphere that is not a "closed system."

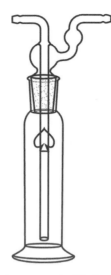

Figure 4.7 A gas drying tower is filled with a porous desiccant.

Figure 4.8 An oil bubbler filled with mineral oil shows that there is a positive pressure of an inert gas in a system and how fast gas is flowing through the system.

Figure 4.9 A gas inlet that can be used to attach a flask to the inert atmosphere manifold.

SAFETY NOTE

It is crucial that reactions not be conducted in closed systems without using extreme precautions. The concept of not *heating* a closed system, as a practical illustration of the gas laws ($PV = nRT$), is impressed on chemistry students at an early stage. However, even without heating, conducting a reaction in a closed system can be hazardous, specifically any reaction that does (or could) produce a gaseous product or by-product. Some common examples include decarboxylation reactions and those involving highly nitrogenous molecules like azides (RN_3) and diazo compounds ($R_2C = N_2$). Application of the gas laws makes this problem readily understood. Recall that a mole of a gas occupies the relatively large volume of 22.4 L. Any reaction that produces a gaseous product in a closed system must substantially increase the pressure because the volume is fixed. Such pressure increases can easily be large enough to separate ground glass joints, even those that are held together with rubberbands, springs, clamps, or even wire. The possibility of spilling reaction mixtures that have their own intrinsic hazards is reason enough not to conduct reactions in closed systems. In a worst-case scenario, the pressure jump may be sufficient to explode the apparatus. This can occur even without flammable reactants, but the consequence is no different—shards of glass and reaction mixture shot around the lab. To attempt to address the possibility of reaction explosion for any cause, safety shields made from a variety of tough, clear plastics are available (Fig. 4.10). These are no substitute for a carefully planned and executed reaction, however.

An integrated manifold combines these functions in one apparatus (Fig. 4.11). A more sophisticated manifold has dual chambers, one of which is attached to the inert gas source and the other is attached to a vacuum source. In this setup a double oblique stopcock enables the lines running to the apparatus to be connected to one

Figure 4.10 A plastic safety shield that can be set up between the reaction and the chemist. Photograph used with permission of Nalge Nunc International Corp. NALGENE® is a registered trademark of Nalge Nunc International Corp.

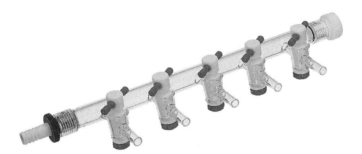

Figure 4.11 A simple manifold to deliver inert gas to several vessels.

manifold or the other. This is a sure method of filling the apparatus with an inert atmosphere: evacuation and then filling. When using this technique, it is essential to have a fairly rapid gas flow into the nitrogen side of the manifold, as evidenced by vigorous bubbling, and ideally a large manifold volume, to avoid sucking air into the manifold when it is exposed (gradually!) to the evacuated flask by rotating the stopcock from one manifold to the other. Of course, when applying vacuum via flexible tubing/hoses, it is essential to use a tubing with rigid walls that will not collapse under reduced pressure. That is, conventional Tygon® tubing will not serve the purpose.

Another feature of Tygon tubing that should be kept in mind is that its flexibility is maintained by a "plasticizer," usually a diester derivative of phthalic acid (e.g., dibutyl phthalate, dioctyl phthalate). These esters are quite soluble in many organic solvents, and in particular chlorinated solvents like dichloromethane. So Tygon tubing must never be used to transfer solvents, and care must be taken when using Tygon tubing to connect gas sources to vessels containing such solvents because the plasticizer will be leached out of the tubing. The two consequences are that the tubing will become somewhat stiff and, more important, that the plasticizer will end up in the reaction mixture and contaminate the product.

Some reactions must be protected from atmospheric moisture but should not be attached to the inert gas manifold. For example, a reaction like a Freidel-Crafts acylation produces HCl, which will likely escape the solution as a gas. HCl should not go into the nitrogen lines. For such situations a classical drying tube may be more appropriate (Fig. 4.12). Desiccants used in a drying tube might include the aforementioned DRIERITE, calcium oxide, barium oxide, or molecular sieves. Plugs of glass wool are used to hold the desiccant in the tube while permitting gas flow. Even when using a drying tube, the apparatus could be filled with dry nitrogen before reaction is begun.

Finally, an alternative to using a manifold to supply inert gas is a balloon (standard latex party balloons work fine, pulled onto a rubber stopper or a short length of rubber tubing). The balloon is generally connected to the apparatus through a gas adapter or even sometimes a needle (despite the obvious incompatibility). This approach is not

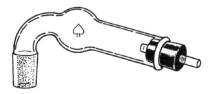

Figure 4.12 A drying tube that can be filled with desiccants such as $CaCl_2$ or molecular sieves.

as effective as a manifold because the volume of gas that can be delivered is much less. Balloons are more useful for delivering reagent gases, such as hydrogen or carbon monoxide, provided that the reaction requires pressures only around atmospheric. Balloons may also be used to provide "makeup" gas to a reagent bottle that makes up for the volume of the liquid that is being withdrawn.

4.5 APPARATUS FOR ADDITION

Facilitating the addition of reagents to a reaction while maintaining an inert atmosphere can be done with several devices. Provided that good inert gas flow is maintained, that the reaction itself is not too sensitive to the atmosphere, and that the reagent can be added all at once, it may be possible to simply remove a stopper, add the reagent, and quickly replace the stopper. Stoppers can be glass or Teflon. Apparatus with Teflon stoppers should not be suddenly exposed to heat, since Teflon expands much faster than glass. Conversely, stuck Teflon stoppers can be freed by immersing them in dry ice/acetone. One drawback of stoppers is that if the reaction experiences a sudden pressure increase, they can become projectiles. More often flexible rubber septa (also called serum caps or serum stoppers) are used for addition of reagents under an inert atmosphere (Fig. 4.13). They fit

Figure 4.13 Rubber septa for some of the most common sizes of ground-glass joints. © Sigma-Aldrich Co. The narrow, hard end is inserted into the joint and the wide, more flexible end is folded down over the outside of the joint.

the internal diameter of the ground glass joint snugly and have a sleeve that is pulled down over the outside of the joint. This friction seal is fine to contain gases but still does not prevent pressure pulses from firing the serum stopper off the flask. Reagents can be added as a solution via a syringe and needle through the serum stopper. Serum stoppers cannot tolerate a great deal of heat, and are not usually dried before use. As organic polymers they are also not very hydrophilic, so the amount of water introduced into reactions by serum stoppers is relatively small, and can only affect small-scale reactions. Serum stoppers can be dramatically swollen by organic solvents such as tetrahydrofuran, toluene, or dichloromethane. Care should be taken in using serum stoppers when they will be exposed to such solvents for long periods, especially at reflux.

The addition on a small scale of a viscous reactant that cannot be measured by volume to a reaction under an inert atmosphere creates a particular challenge. The best approach is to prepare and transfer a solution of the compound in the reaction solvent. This can be done by evaporating a solution of the compound into a tared 5 to 10 mL pear-shaped flask, the type that comes to a sharp point (maximizing the volume of solution that can be drawn into a syringe needle). The flask is fitted with a serum stopper and placed under an inert atmosphere. A minimum volume of a dry solvent is added (typically by washing it down the walls of the flask) and the compound is dissolved. Another syringe is used to transfer the solution into the reaction flask. Half of the original volume of solvent is added to the pear-shaped flask and the residue is dissolved and transferred. The latter process is repeated.

A classical method of addition of liquids to a reaction is an addition funnel. Within a closed system it is necessary that the funnel be of the pressure equalizing type (Fig. 4.14), with a tube connecting the space above the reagent to the space above the addition tube but below the joint. A potential problem with addition funnels is that the addition rate is dependent not only on the position of the stopcock but on the hydrostatic pressure of the solvent, meaning that as liquid is depleted, the addition rate slows. Maintaining a constant rate of addition using an addition funnel requires close monitoring that is tedious. Another important method of liquid addition is a syringe pump (Fig. 4.15), also called (often in experimental descriptions) a motor-driven syringe. This instrument works with a standard

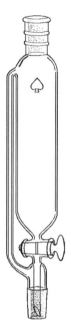

Figure 4.14 A pressure equalizing dropping funnel for addition under an inert atmosphere.

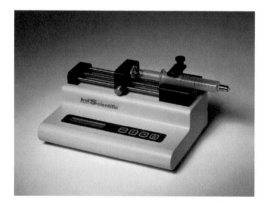

Figure 4.15 A syringe pump accepts a variety of sizes of syringes and drives a ram against the plunger at a constant rate. The rate of liquid delivery is variable based on the size of the syringe and the setting of the syringe pump.

syringe (see Section 8.4), basically clamping it in place and driving a ram against the plunger. The syringe pump can be set for a range of delivery rates for a given syringe size, and it is a superior tool to obtain a constant rate of addition.

It is worth specifically mentioning the purpose of varying the addition rate in a reaction. One obvious purpose is that the reaction is exothermic, and fast addition rates lead to thermal runaway. That is, the heat of reaction from addition of the first portion of the reagent raises the temperature of the reaction mixture such that the next portion of the reagent reacts even faster, creating even more heat that raises the reaction temperature even more.

Slow addition of reagents is commonly used to minimize the reaction of a reagent with itself. The principle is as follows. We wish A to react with B to form the product P, but A can also react with itself. From the rate laws for each reaction, the relative rates of the two processes can be compared (Fig. 4.16). This analysis shows that the way to favor reaction with B (i.e., $rate_2$) is to keep the concentration of A as low as possible. This is true irrespective of the relative magnitude of the rate constants for the two reactions or the B concentration. A situation where the A concentration is as low as possible is easy to arrange by adding A slowly to the solution of B. All of the A that is added in each drop is ideally consumed by reaction with B before the next drop of A is added.

Slow addition to keep the reagent concentration low is also used to favor intramolecular reactions at the expense of intermolecular reactions. For example, the formation of large rings (Fig. 4.17, $n = 7$ or greater) must compete with the formation of oligomeric

$$\text{A} \longrightarrow \begin{cases} \text{A} \quad \text{A-A} \quad rate_1 = k_1 [A]^2 \\ \text{B} \quad \text{P} \quad rate_2 = k_2 [A][B] \end{cases} \qquad \frac{rate_2}{rate_1} = \frac{k_2 [B]}{k_1 [A]}$$

$$\therefore, \text{ as } [A] \rightarrow 0, rate_2/rate_1 \rightarrow \infty$$

Figure 4.16 A kinetic scheme that demonstrates how the mode of addition can affect the proportion of self-coupling versus cross-coupling in a reaction involving reagents A and B.

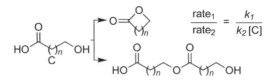

Figure 4.17 A kinetic analysis demonstrates that slow addition, maintaining a low concentration of reagent C, maximizes the proportion of intramolecular reaction to intermolecular reactions.

Figure 4.18 A powder addition funnel can be used only with solids that do not adhere to glass. It also requires a relatively large-scale reaction.

derivatives. Writing the rate law for each reaction, taking their ratio, and solving the equation is an exercise left to the reader. This analysis shows that the way to favor intramolecular reaction (i.e., $rate_1$) is to keep the concentration of C as low as possible, which again can be achieved by a slow dropwise addition to the reaction mixture.

A powder addition funnel can be used to add free-flowing solids (Fig. 4.18). Alternatively, tip flasks (with a bent neck) retain solids when in the down position and allow them to drop into the reaction mixture when rotated into the up position.

These techniques for working under an inert atmosphere will find the most use in ordinary organic laboratories. Much more sophisticated methods are also available, as reflected in the outstanding text of Shriver (Shriver and Drezdzon, 1986).

4.6   CONDENSERS

When reactions are conducted under reflux, a condenser is used. Two basic types of condensers are available: water and dry ice (Figs. 4.19, 4.20). The dry ice condenser is used for cryogenic solvents like liquid ammonia. It is fitted to the flask and filled with dry ice/iso-propanol. The chemist must attend this condenser closely to ensure that the dry ice is replenished when needed. Water condensers should be connected to a supply of cold water with Tygon tubing. It can greatly enhance the ability to connect varying combinations of condensers into the system if the waterline is fitted with Nalgene® connectors. Each condenser is equipped with an inlet and outlet tube with a connector, allowing the easy insertion of a condenser into the system. The water flow in condensers should be carefully controlled with a needle valve or pressure regulator if reactions are to be left unattended. Otherwise, the drop in pressure when water use rises may cause the reaction to lose cooling, or a rise in water pressure

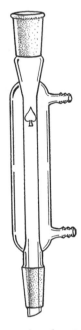

Figure 4.19 A conventional water-cooled condenser.

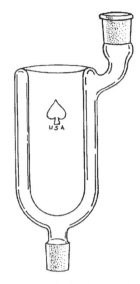

Figure 4.20 A dry ice condenser.

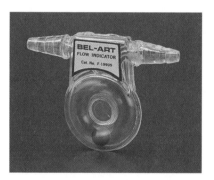

Figure 4.21 A flow indicator for insertion into tubing that delivers cooling water to a condenser. The ball circles the device when water is flowing through the tubing. Bel-Art Products.

could cause a connection to loosen and lead to a flood. It is also a good idea to insert a flow indicator (Fig. 4.21) into the water cooling lines so that the speed of water flow is apparent. Floods from cooling water overflows are one of the most common laboratory accidents. While not as severe as some chemistry laboratory accidents, floods are still quite damaging, and the water damage can extend from the lab of the experimenter to labs in floors below. It is essential for this reason that all connections of water hoses to hose connectors on glassware be secured with copper wire.

4.7   OTHER EQUIPMENT AND CONSIDERATIONS

A jack stand or lab jack (Fig. 4.22), essentially a small scissors-type jack, is extremely useful for a reaction setup. It permits heating/ cooling baths, magnetic stirrers, and/or hotplates to be brought up to (and quickly withdrawn from) a flask that is mounted at a fixed height to a fixed grid of metal rods. These grids are standard in most laboratory furnishings. Ring stands are not recommended for supporting reaction flasks because they are relatively unstable and easily toppled. The jack stand is also used to adjust the distance between a magnetic stirring motor and the stir bar in the flask to achieve optimal stirring.

Minimizing exposure of laboratory workers to chemicals is one of the hallmarks of laboratory safety. There is little justification for conducting reactions anyplace except in the fume hood. Most modern laboratories are set up with this in mind, with areas outside the fume hood reserved for other operations. One example of how

Figure 4.22 A lab-jack or jack stand is very useful for raising and lowering temperature control baths to a flask that is held at a fixed position.

this principle can be important is with reagents that fall into the class of "sensitizers." Often these are strong alkylating agents. After an initial exposure of the chemist to such a reagent, future exposures to that compound may promote an allergic reaction like a rash or an anaphylactic reaction like asthma, and this sensitivity may extend to other compounds in that class. An initial exposure to one compound may then prevent the chemist from ever again working with a whole range of reagents. Knowledge aforethought of this phenomenon can inform many laboratory procedures. For example, it is fairly easy to become sensitized to methylating agents. Knowing this, it would be unwise to quench a reaction involving excess tosyl chloride with methanol, which would be converted to methyl tosylate (eq. 10). A better choice for a quenching agent is ethanol, which is almost as reactive as methanol in quenching the tosyl chloride but generates ethyl tosylate, which is not nearly as reactive as an alkylating agent as methyl tosylate.

$$\text{MeOH} \quad + \quad \text{TsCl} \quad \longrightarrow \quad \text{Me-OTs} \qquad (10)$$

CHAPTER
5
TEMPERATURE CONTROL

Reaction temperatures commonly used in organic reactions range from −100°C to nearly 200°C. For reactions below room temperature, an alcohol thermometer will typically be used. For reactions above room temperature, a mercury thermometer should be used. Thermometer adapters made of Teflon (Fig. 5.1) or glass are available for standard sizes of ground glass joints. The former have neoprene rings that can be tightened onto the thermometer via a screw fitting. The latter might work in this way, or have very small tapered ground glass joints that accept thermometers with matching ground glass fittings. The latter thermometers are less attractive because they are more expensive, and because their depth of immersion into the flask, which may not be correct for every reaction, is fixed. A tricky issue in getting a workable reaction setup is choosing the right flask and solvent volumes to allow the thermometer bulb to be immersed in the solution without being struck by the magnetic stirring bar, and to have this situation persist once the solution is being stirred, which often creates a vortex in the center of the solution just where the thermometer bulb sits.

5.1 HEATING

For reactions above room temperature, oil baths provide the best control. Silicone oil, also called DOWTHERM™ fluid, heat transfer fluid, or just transfer fluid, is preferred. Many variants with different properties are available—they typically have a working temperature up to about 250°C (Fig. 5.2). Do not use paraffin, mineral, or white oils, since they smoke and discolor and have a low flashpoint. Oil baths are best heated electrically by an integral heating element (Fig. 5.3) controlled by a Variac or Powerstat. Some labs have a tradition of using home-made oil bath heaters consisting of coils of nichrome

The Synthetic Organic Chemist's Companion, by Michael C. Pirrung
Copyright © 2007 John Wiley & Sons, Inc.

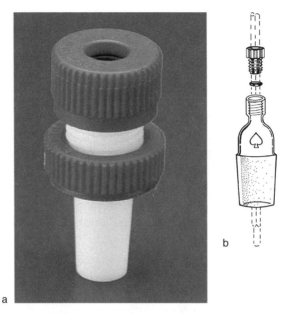

Figure 5.1 (*a*) A Teflon thermometer adapter fits into a standard ground-glass joint and accepts a glass thermometer or gas inlet tube. The seal is made by an O-ring that is compressed against the thermometer or tube by screwing in the knurled knob. Bel-Art Products. (*b*) Thermometer adapters can also be made from glass.

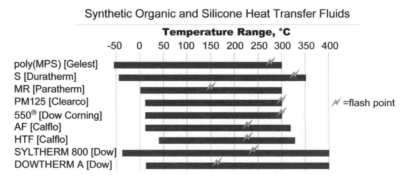

Figure 5.2 The working temperature ranges and flash points for several commercial heat transfer fluids.

Figure 5.3 An oil bath with an integral heating element.

wire connected to a Variac or Powerstat with patch cords. These setups typically have exposed electrical contacts and are not safe.

Oil baths might also be heated by a hotplate or stirring hotplate. This is probably the easiest and most common way to heat a bath. The temperature the bath can reach is limited by poor thermal contact between the top of the hotplate and the bath container, so do not use foil or insulating material on the top plate. Because the temperature of the hotplate must be higher, sometimes much higher, than the temperature of the bath, it presents a fire hazard when hotter than the flash point of the oil. Few hotplates are explosion-proof. Spills of oil or organic liquids on the hot surface can also lead to a fire, or at least discoloration or failure of the ceramic top. The excess heat given off by the hotplate can create problems by heating the surrounding apparatus. It is important to use a thermometer in the bath itself so that its temperature can be directly monitored and controlled. A magnetic stir bar should also be used in the oil bath to ensure that the temperature is uniform. Electronic devices to monitor and maintain the temperature are available—one example is the THERM-O-WATCH® (Fig. 5.4). It has a sensor that clamps onto a mercury thermometer to measure the mercury level, and turns the heat on and off as needed to maintain the level and therefore the temperature. Of course, this cycling can result in drifting of the bath temperature by ±10°C. When filling an oil bath, be sure to take into account the volume displaced by the flask and volume expansion upon heating.

Heating mantles are really only satisfactory for heating reactions conducted at reflux, and even then there is a risk of uneven heat

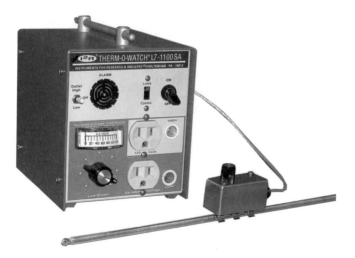

Figure 5.4 A THERM-O-WATCH monitors the temperature via a sensor placed on a mercury thermometer and controls the power being delivered to a heating device to maintain a constant temperature. Similar sensors are available to ensure that water is flowing before heat is applied to a system that requires cooling.

distribution across the mantle. Take care that reflux is gentle, ensuring that the pot is not appreciably above reflux temperatures. It can be difficult to get magnetic stirring to work through a heating mantle, however. Heating mantles may be used for solvent stills. Steam baths may be used to heat reactions to temperatures in the 70°C range but are hardly available in the modern organic chemistry laboratory.

A modern approach to the heating of organic reaction mixtures uses microwaves. Specialized apparatus made for the laboratory is highly preferred over conventional domestic microwave ovens, which have hazards and other disadvantages. Commercial microwave reactors provide capabilities to monitor, control, and maintain the temperature and/or microwave power and conduct reactions above atmospheric pressure. Microwave-based reactions typically use a polar organic solvent to absorb the microwaves. The main advantage of microwave heating for reactions seems to be the speed with which the reaction mixture reaches temperature and with which it can be cooled following reactions. To some extent the best analogy of microwave heating to conventional organic reaction techniques may be flow pyrolysis, a method that has a good history in organic synthesis but is operationally difficult to implement and therefore not discussed here. Microwave chemistry has become a prominent subfield of synthetic organic chemistry, and texts are available that describe it in great detail (Kappe and Stadler, 2005).

5.2 COOLING

For low temperature several different liquid baths are used with the few available coolants. Temperatures of various baths are given in Table 1. However, temperatures around the freezing points of the coolants themselves are far easier to maintain. At the other temperatures, one is often relying on the freezing point of the liquid to buffer the effect of added coolant. This process seems to degenerate into incessant addition of coolant and then liquid, which is not optimal, since it is certain that the reaction temperature is actually fluctuating around the melting point of the liquid (not to mention the tedium imposed on the chemist). This problem may be addressed by using eutectic ice/salt mixtures of specific proportions. When temperature maintenance for long periods is needed, home-made insulated baths consisting of one crystallizing dish inside another, with the space between filled with vermiculite and sealed with silicone sealer (Fig. 5.5), are sometimes used to conserve coolant. This is a cheap alternative to a dewar dish (Fig. 5.6), and also gets around the problem that many dewars prevent effective operation of magnetic stirrers.

Liquid nitrogen is one of the few cryogenic liquids routinely used in the organic chemistry laboratory, and special considerations attend its use. Obviously it can instantly freeze flesh, so it must be used with caution and with appropriate personal protective equipment. "LN" is transferred from large storage dewars (Fig. 5.7) into smaller dewars (Fig. 5.8) by simply opening the outlet of the former. Cold nitrogen gas will initially flow into the receiver, to be followed by liquid once the transfer line has been cooled below the boiling point of nitrogen. Then liquid reaching the receiver will be vaporized in cooling it below the boiling point of nitrogen, and finally liquid will begin to accumulate in the receiver. This process repeats with every transfer into a new container.

Chillers offered by several different manufacturers (Fig. 5.9) use a cooling probe in place of a solid coolant and, most usefully, have

Table 1 Coolant/Liquid Combinations to Achieve Specified Coolant Bath Temperatures

Temperature	Coolant/Liquid
0°C	Ice/H_2O
−10 to −15°C	Ice/acetone
−10 to −15°C	100 g Ice/33 g NaCl
−16°C	100 g Ice/25 g NH_4Cl
−28°C	100 g Ice/67 g NaBr
−34°C	100 g Ice/84 g $MgCl_2 \cdot 6H_2O$
−55°C	100 g Ice/143 g $CaCl_2 \cdot 6H_2O$
−10.5°C	Dry ice/ethylene glycol
−12°C	Dry ice/cycloheptane
−15°C	Dry ice/benzyl alcohol
−25°C	Dry ice/1,3-dichlorobenezene
−29°C	Dry ice/o-xylene
−30 to −45°C	Dry ice/aq. $CaCl_2$ of varying concentrations
−32°C	Dry ice/m-toluidine
−38°C	Dry ice/3-heptanone
−41°C	Dry ice/acetonitrile
−46°C	Dry ice/cyclohexanone
−47°C	Dry ice/m-xylene
−56°C	Dry ice/n-octane
−78°C	Dry ice/commercial isopropanol*
−83.6°C	Liquid N_2/ethyl acetate
−89°C	Liquid N_2/n-butanol
−94°C	Liquid N_2/hexane
−94.6°C	Liquid N_2/acetone
−95.1°C	Liquid N_2/toluene
−98°C	Liquid N_2/methanol
−104°C	Liquid N_2/cyclohexane
−116°C	Liquid N_2/ethanol
−120°C	Liquid N_2/4:1:1 petroleum ether/isopropanol/acetone
−131°C	Liquid N_2/n-pentane
−160°C	Liquid N_2/isopentane

*Dry ice/isopropanol is recommended over dry ice/acetone because it fizzes less, is less volatile, and is less flammable.

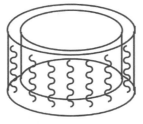

Figure 5.5 A homemade insulated cold bath.

Figure 5.6 Low-profile dewar flasks for use as cold baths.

Figure 5.7 Large dewars for storage of liquid nitrogen. These tanks are around 1.5 m in height. Photo courtesy of Taylor-Wharton.

Figure 5.8 Smaller dewar container used to transfer liquid nitrogen from the central supply dewar to the individual lab. © Sigma-Aldrich Co.

Figure 5.9 A chiller with a probe that can be used in place of a coolant. Photo provided by Brinkmann.

a temperature probe so that a thermostat can maintain the temperature at a set-point. These chillers can be essential to maintaining reactions at low temperature for longer than the chemist's stamina and patience. A cooling liquid commensurate with the desired temperature must be used.

CHAPTER

6

SOLVENTS

6.1 SELECTION

The selection of a solvent for today's reaction may be obvious because a procedure is available for this or a comparable reaction. A detailed consideration of solvent effects on organic reactions is not intended here; excellent reference texts are available (Reichardt, 2002). In addition there are some statistical methods that attempt to segregate solvents based on their aggregate molecular properties (Carlson, Lunstedt, and Albano, 1985). A chart of these solvent properties is supplied in Appendix 2 as well as on the foldout from the inside back cover. Be sure to consult tables of freezing points of potential reaction solvents before choosing a low temperature for a reaction. These types of missteps are often made with solvents like benzene, DMSO, and acetic acid that have melting points just below room temperature. Solvent mixtures have lower freezing points and may allow access to lower temperatures. For example, 4:4:1 tetrahydrofuran/ethyl ether/pentane (the so-called Trapp solvent) is useful to −120°C.

In choosing a solvent, knowledge of the risk of exposure of the chemist to the solvent must be taken into account. The relative risk related to particular solvents, and therefore the precautions that should be observed, may not always be obvious. Precautions appropriate to the particular solvent, such as work in the hood or ventilator use for inhalation toxicity, and gloves, lab coats, and other personal protective equipment for skin toxicity, should be observed.

6.2 PURITY

As the component of the reaction mixture present in the largest amount, the purity of the solvent can have a major impact on a reaction. Of particular concern is the removal of dissolved water or

The Synthetic Organic Chemist's Companion, by Michael C. Pirrung
Copyright © 2007 John Wiley & Sons, Inc.

other protic substances from solvents. Water is pervasive and there-fore will be present in any solvent with an affinity for it. This includes essentially all ethers and dipolar aprotic solvents; it is less of a problem with hydrocarbons. Not only will dissolved water serve as an acid for any of the interesting organic anions frequently used in synthesis (eq. 11), water interferes with the formation of organome-tallic reagents like Grignards from alkyl halides. Solvent purification typically has involved distillation from a drying agent appropriate for each solvent. A classical choice for ethers has been sodium metal, often in the presence of an indicator like benzophenone. Sodium's reducing properties address the very real concern with peroxides that can be formed from ethers in the presence of oxygen.

$$Li\diagdown CN \; + \quad H_2O \; \longrightarrow \quad CH_3CN \quad + \quad LiOH \qquad (11)$$

A surprisingly large number of organic compounds react spon-taneously with O_2 in the air to form peroxides. Butadiene (and likely other dienes) and isopropyl ether (and likely other ethers with tertiary α-hydrogens) can form explosive levels of peroxides even without concentration by evaporation or distillation. A large number of compounds can form explosive levels of peroxides upon concentration. They include acetaldehyde, diacetylene, benzyl alcohol, 2-butanol, cumene, cyclohexanol, cyclohexene, 2-cyclohexen-1-ol, decahydronaphthalene, dicyclopentadiene, die-thyl ether, diglyme, dioxanes, glyme, 2-hexanol, 4-heptanol, methylacetylene, 3-methyl-1-butanol, methylcyclopentane, methyl isobutyl ketone, 4-methyl-2-pentanol, 2-pentanol, 4-penten-1-ol, 1-phenylethanol, 2-phenylethanol, 2-propanol, tetrahydrofuran, and tetrahydronaphthalene.

Peroxide-forming chemicals should be stored in the original man-ufacturer's container whenever possible. This is very important in the case of diethyl ether because the iron in the steel containers in which it is shipped acts as a peroxide inhibitor. In general, peroxide-forming chemicals should be stored in sealed, air-impermeable con-tainers and should be kept away from light, which can initiate peroxide formation. Dark amber glass with a tight-fitting cap is rec-ommended. Peroxide-forming chemicals can be stored for 3 months after opening the container if they form peroxides without concen-tration, and for 12 months after opening the containers if they form peroxides with concentration. Materials may be retained beyond this suggested shelf life only if they have been tested for peroxides (see below), show peroxide concentrations lower than 100 ppm, and are re-tested frequently.

All solvents that are to be distilled should be tested for the presence of peroxides regardless of their age. A safe level for peroxides is considered to be less than 100ppm. While several methods are available to test for peroxides in the laboratory, the most convenient is the use of peroxide test strips available from many chemical suppliers. For volatile organic chemicals, the test strip is immersed in the chemical for one second. The chemist breathes on the strip for 15 to 30 seconds or until the color stabilizes, and the color is compared with a provided colorimetric scale. Any container found to have a peroxide concentration at or over 100 ppm should be disposed of (with the assistance of the safety office).

SAFETY NOTE

Researchers should never test containers of unknown age or origin for peroxides. Older containers are far more likely to have concentrated peroxides or peroxide crystallization in the cap threads. Therefore they can present a serious hazard when opened for testing.

Solvents stills are traditionally a continually maintained fixture in many organic synthesis laboratories. However, still pots containing highly flammable solvents and reactive metals pose a significant fire risk, not only during use but in quenching when the still must be regenerated. This *Companion* will therefore not provide a detailed procedure for setting up a sodium solvent still. It makes more sense to maintain a still system continuously when other, less dangerous drying agents are used and the lab uses that solvent regularly. An apparatus for that purpose collects distilled solvent in a bulb (Fig. 6.1). Stills are usually maintained at a low heating level (not enough to cause reflux) so that they quickly come to reflux to supply needed solvent. After collecting solvent from a still and turning the heat back down, the nitrogen flow must be maintained at a sufficient rate

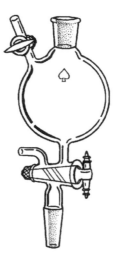

Figure 6.1 A still head for a solvent still.

so as to avoid pulling oil back into the system from the bubbler (recall the gas laws).

Available drying agents other than sodium include those that react with water chemically (eqs. 12, 13, 14). The desiccant(s) recommended for some commonly used solvents are given in Table 2. Drying agents that sequester water, such as molecular sieves, are useful with essentially any solvent given in Table 2. Sieves are zeolites (alkali aluminum silicates) with specific pore sizes; for example, 3 Å molecular sieves permit only water to penetrate into the solid. Molecular sieves are amazingly effective and broadly useful desiccants provided they have been activated, which is done by heating them under vacuum, ideally to 300–350°C, but heating to 150–200°C is still effective. They must be stored under an anhydrous atmosphere, like any desiccant. The properties of several classes of molecular sieves are given in Table 3. They are available in a variety of physical forms, including beads, pellets, and powders.

$$H_2O \quad + \quad P_2O_5 \quad \longrightarrow \qquad \underset{HO}{\overset{O}{\underset{}{\parallel}}} P \underset{O}{\overset{O}{\underset{}{\parallel}}} O \underset{}{\overset{O}{\underset{}{\parallel}}} P \underset{OH}{\overset{O}{\underset{}{\parallel}}} \qquad (12)$$

$$H_2O \quad + \quad CaH_2 \quad \longrightarrow \quad CaO \quad + \quad 2\,H_2 \qquad (13)$$

$$H_2O \quad + \quad Ac_2O \quad \longrightarrow \qquad 2\,HOAc \qquad (14)$$

Table 2 Desiccant(s) and Purification Method for Some Common Solvents

Solvent	Drying Agents	Fraction/Pressure/Other Info
Acetone	$CaSO_4$	
Acetonitrile	P_2O_5, CaH_2	
tert-Butanol	CaH_2, Al/Hg, $Mg(OEt)_2$	Melting point of 25°C means solid may block condenser
Dichloromethane	P_2O_5, CaH_2	Or stand over molecular sieves (24 h)
Diisopropylamine	CaH_2	
Dimethylsulfoxide	CaH_2	Stand over CaH_2 overnight, distill *in vacuo** (72°C, 12 torr)
Dimethylformamide	BaO, P_2O_5	Distill *in vacuo** (55°C, 20 torr) (decomposes at boiling point at atmospheric pressure)
Ethanol	$Mg(OEt)_2$, CaH_2	From Mg metal
Ethyl acetate	P_2O_5, Ac_2O, K_2CO_3	Collect all distillate
Hexanes	CaH_2	Collect all distillate
Hexamethylphosphoramide	CaH_2	Distill *in vacuo** (115°C, 15 torr)
Nitromethane	mol sieves, $CaSO_4$	Collect all distillate
N-Methylpyrrolidone	BaO, CaH_2	Distill *in vacuo** (96°C, 24 torr)
Methanol	$Mg(OMe)_2$, CaH_2	From Mg metal
Petroleum ether	CaH_2	Collect all distilling below 55°C
Pentane	CaH_2	Collect all distilling below 55°C
Pyridine	BaO, CaH_2	Collect all distillate
Toluene	CaH_2, P_2O_5	Collect all distillate
Trichloroethylene	K_2CO_3	Collect all distillate
Triethylamine	CaH_2	Collect all distillate

*Aspirator vacuum is usually effective. Be certain to insert a drying tube into the vacuum line.

Table 3 **Types of Molecular Sieves**

Type	Pore Size (Å)	Description
3A	3	Absorbs H_2O; good for drying polar liquids.
4A	4	Absorbs H_2O, C_2H_5OH, C_2H_4, C_2H_6, C_3H_6 (but not higher hydrocarbons); good for drying nonpolar liquids and gases.
5A	5	Absorbs n-C_4H_9OH and n-C_4H_{10}, but not iso-alkanes or rings ≥C4.
10X	8	Absorbs highly branched hydrocarbons and aromatics; used for purification and drying of gases.
13X	10	Particularly good for drying hexamethylphosphoramide, $[(CH_3)_2N]_3PO$.

A superior solution to the problem of generating anhydrous solvents was provided by Grubbs (Pangborn et al., 1996). His method relies on a source of purified bulk solvents that is suited to using cartridges of highly activated alumina as the desiccant. Cartridges of supported copper catalyst are optionally used for removal of dissolved oxygen from hydrocarbons. Complete systems that enable the drying, manipulation and supply of these solvents under inert gas are available (Fig. 6.2). Solvents can even be piped from the system directly into inert atmosphere glove boxes. The range of solvents that can be purified is wide, including acetonitrile, methanol, DMF, ethyl ether, tetrahydrofuran, hexanes, dimethoxyethane, toluene, pyridine, triethylamine, DMSO, and dichloromethane. Nitromethane cannot be used with these systems, and for alcohols, a glass insert in the cartridge is necessary to prevent leaching of desiccant into the solvent.

6.3 DEGASSING

It may be important for some reactions to remove dissolved oxygen from the solvent. If this is not intrinsic to the solvent purification method itself (see above), solvent degassing can be performed in one of two ways. The rigorous procedure is called freeze-pump-thaw. The solvent is frozen well below its freezing point with a coolant,

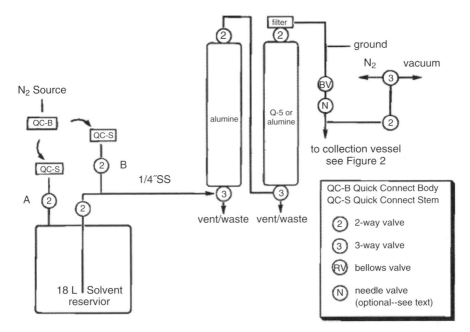

Figure 6.2 A schematic of "Grubbs" solvent purification system. Degassing: Pressurize at B, vent to hood from A. Purging tubing with N_2: Pressurize at B, vent to the three-way valve. Purification/collection: Pressurize at A. Reprinted with permission from Pangborn et al., 1996. Copyright 1996, American Chemical Society.

which for some solvents could be dry ice/isopropanol, and for others could be liquid nitrogen. A reasonably good vacuum (ca. 1 torr) is applied for several minutes, and the container is refilled with inert gas. The solvent is thawed, and this cycle is repeated two to three times. The easier procedure that most synthetic chemists will use is to bubble an inert gas through the solvent for about 15 minutes using a fritted gas dispersion tube (Fig. 6.3). This method may not work well with particularly volatile solvents, which will evaporate.

Each solvent has its own unique properties that the chemist using it routinely should become familiar with. For example, chloroform has a tendency to undergo a slow decomposition to give HCl. This process is inhibited by the presence of ethanol, which is why commercial chloroform often includes low percentages of ethanol as a stabilizer. The potential presence of ethanol or HCl in chloroform must therefore always be considered.

Figure 6.3 A gas dispersion or sparging tube has a glass frit at the end that creates voluminous bubbles.

6.4   AMMONIA

A very special example of a reaction solvent is liquid ammonia. It is most commonly used for dissolving metal reductions. Liquid ammonia is obtained in a tank (see Section 3.2) that is connected to an apparatus bearing a charged dry ice condenser and a flask cooled in a −78°C bath. The tank valve is opened, ammonia is condensed into the flask, and the ammonia line is replaced by an inert gas line. The ammonia in the tank is not necessarily anhydrous, but for many reactions where excess sodium may be used, for example, cleavage of a benzyl ether using sodium in ammonia (eq. 15), this may not be a problem.

$$Ph\overset{}{\diagup}O\overset{}{\diagdown}\overset{}{\diagup}\overset{}{\diagdown}CO_2Na \xrightarrow[NH_3\ (l)]{Na^\circ} HO\overset{}{\diagdown}\overset{}{\diagup}\overset{}{\diagdown}CO_2Na \quad (15)$$

Sodium is packaged in a variety of ways and often appears as a large ingot, caked with NaOH from reaction with moisture in the atmosphere. Immediately upon opening the packaging, the sodium should be placed under mineral oil or a high-boiling hydrocarbon solvent like xylene to protect it from the atmosphere. The ductile metal is cut into small chunks (maximum 2 cm square) with a knife or spatula while under the mineral oil. The metal is transferred with forceps to a container of toluene, which is used to remove the oil, and then to a container of the solvent, which is used to remove the toluene.

Small pieces of sodium are added to the cooled ammonia. Sodium reacts with any water present to form sodium hydroxide (eq. 16). Once all the water has been consumed, addition of more sodium will result in the formation of the unmistakable blue solution of solvated electrons (eq. 17). The stoichiometric quantity of sodium needed for the reaction can now be added.

Alternatively, if the reaction in question requires that *no* hydroxide be present (e.g., a reductive alkylation of an enone; eq. 18), the ammonia can first be condensed into a cooled flask to which sodium is added until the blue color persists. This flask is then connected to the gas inlet of the dry ice condenser and anhydrous ammonia is distilled into the reaction apparatus by slightly and carefully warming the flask with a water bath. Another approach to obtaining anhydrous ammonia is to pass it through two drying towers containing KOH and CaO before being condensed into the apparatus.

$$Na^\circ \quad + \quad H_2O \quad \longrightarrow \quad H_2 \quad + \quad NaOH \quad (16)$$

$$Na^\circ \quad + \quad NH_3 \quad \longrightarrow \quad e^-(NH_3)_4\ Na^\oplus \quad (17)$$

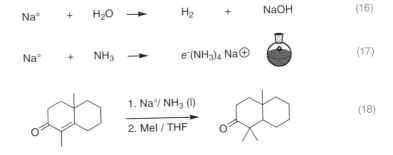

$$(18)$$

CHAPTER

7

THE RESEARCH NOTEBOOK

The keeping of a proper research notebook is an essential part of doing any kind of science. The training in this skill that most students receive in the organic teaching laboratory is rarely adequate for the research laboratory setting. This is especially true for industrial chemistry, where the research notebook is in many cases the first documentation of the conception and/or reduction to practice of a chemical idea. Because the date of conception is still an important facet of the patenting process in the United States, it is essential that the research notebook adequately describe the idea, that the date be clearly identified, and that the notebook page(s) be witnessed by a scientist qualified to understand the chemical concepts involved. Patent litigation has literally hinged on a particular compound being prepared in two different labs in two different companies on consecutive days.

While the foregoing level of dedication to procedure is unlikely to be observed in the academic setting, instituting good habits in the keeping of a research notebook while in training will make the transition to the industrial setting that much easier for the chemist. Chemists who are attempting to repeat reactions done by earlier members of a research group will rapidly come to appreciate their predecessors who provided detailed procedures, and to scorn those who provided sparse details of how they actually got that key compound purified or obtained that excellent yield. The experience of trying to replicate an experiment from another person's notebook would be excellent training for any novice chemist. It would also likely provide an understanding and appreciation for the necessity to adequately document one's reactions in the notebook.

For synthetic reactions, each research notebook page will generally have a specified set of elements. Each new reaction should begin on a new page. A structural equation describing the reaction or a structure of the compound under study should be at the top of the

The Synthetic Organic Chemist's Companion, by Michael C. Pirrung

page for easy visual retrieval. Preparation of a *complete* table of reagents and products (with molecular weights, masses, moles, and mole ratio for each, plus other relevant physical properties like density or boiling point) is a lesson many students have learned in the teaching lab. Too often it is the only element consistently found on notebook pages. This table not only helps ensure that reagents are used in the proportions intended, but it also facilitates calculations of the theoretical yield or the volumes of reagents to use. If a literature reference or references are being followed in the execution of this reaction, citations should be provided. If the apparatus used for a reaction is at all exotic or unusual, drawings or pictures should be provided. Considering the growing prevalence of digital cameras, the latter may become much more common. A narrative of how the reaction was actually conducted should be handwritten in the notebook while the experiment is being performed. Any specific experimental observations (color changes, exotherms, bubbling, etc.) should also be provided.

For characterizing the outcome of reactions, there are typically several different types of data that are not amenable to inscription in the notebook, including spectral or chromatographic plots. For each spectrum or piece of data that is collected but not placed directly in the notebook, a unique identifying code (e.g., a notebook identifier [often the chemist's initials along with the notebook number], the page number, and the spectrum number) should be written in the book, along with a very brief indication of what the data show. The same identifier should be placed on the spectra, plots, and other exhibits. These data can be collected in a companion binder that carries an identifier that links it uniquely to that specific notebook.

In order to make it as easy as possible to locate specific experiments in each notebook, the first 10 to 15 pages should be left blank. This area can be used to maintain a Table of Contents. Like the reaction equation at the top of each page, table entries should be structural equations for easy visual retrieval. When a reaction is repeated, a page number should be added to the equation that already exists in the table. This way it is simple to locate all of the trials of a particular reaction over time and observe the ways in which it was conducted. Finally, the Table of Contents must be kept up to date.

An example of one page of a research notebook prepared by an early graduate student in the author's lab is provided (Fig. 7.1). It is a model of how notebooks should be kept. Alas, the author cannot take credit for training this chemist in keeping a notebook, because before beginning graduate school he had already worked in industry,

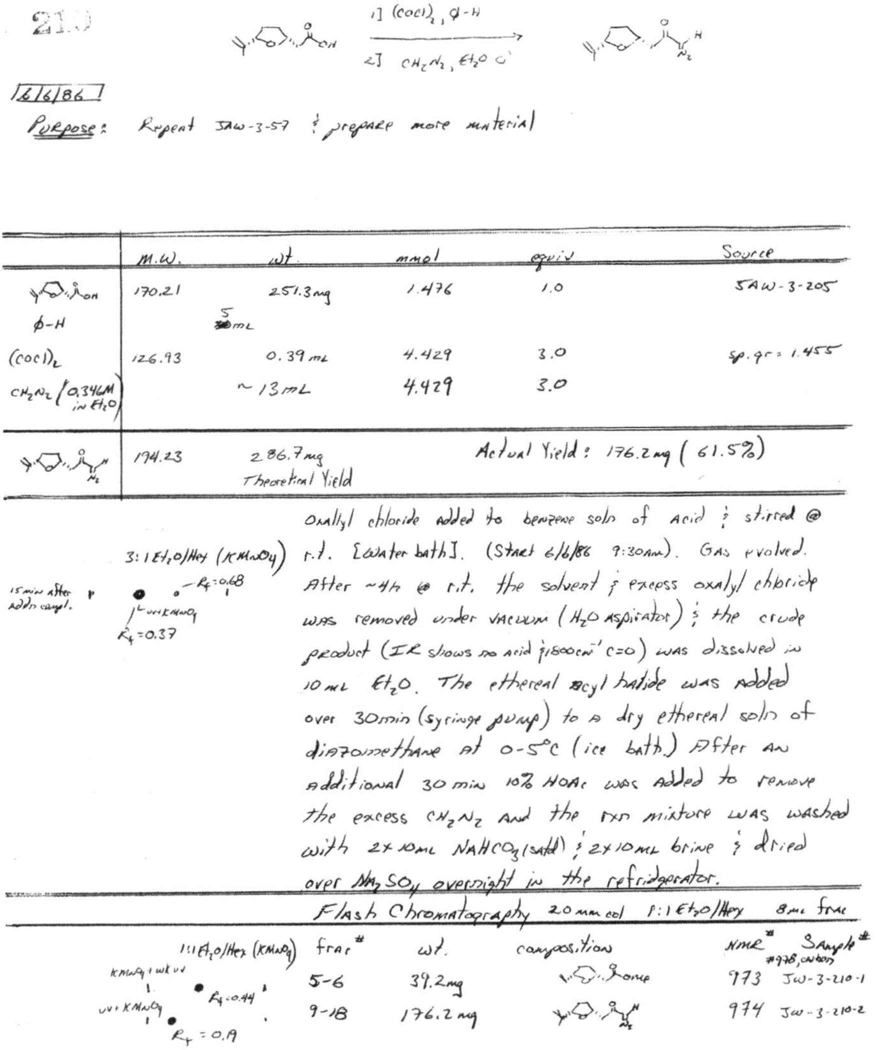

Figure 7.1 A page from the research notebook of John A. Werner.

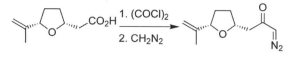

Figure 7.2 The reaction conducted by Werner.

where their much higher standards had already been instilled in him. The reaction that he was conducting is more easily seen in Figure 7.2. Note that he dated the page and referred to an earlier page in the notebook with another preparation of this compound. He provided the source (in this case, an earlier reaction) of his starting acid. He had available a solution of diazomethane of known concentration (though its source is not mentioned—a minor omission). He drew a picture of the TLC of the reaction mixture, specifying the eluting solvent and the stain used. Potassium permanganate was useful in this case because the reactant and product include an alkene. He recorded a starting time for the reaction and noted the gas evolution (expected in the conversion of an acid to an acid chloride with oxalyl chloride). When he isolated the product of the first step, he obtained an IR spectrum to demonstrate conversion of the acid to the acid chloride. For this reaction these data are far more informative than TLC, as the acid chloride would simply hydrolyze on the silica gel, returning the acid and suggesting that no reaction occurred. He described that he added the acid chloride to the diazomethane (it is necessary that diazomethane be in excess to obtain the desired α-diazoketone, and as described elsewhere this can be accomplished by slow addition). He described how addition was performed and over what time. He used acetic acid to quench the reaction, which reacts with any remaining diazomethane to form methyl acetate, which can be easily removed by evaporation. He described the phase partitioning workup that would remove any free acids (starting material or acetic acid), and that he dried the reaction mixture with sodium sulfate (over night). The diazoketone might be sensitive to acidic drying agents like magnesium sulfate. He purified the crude product by flash chromatography and specified the eluting solvent, the size of the column, and the size of the fractions. Note that his solvent for this preparative column is less polar than used for his analytical TLC. He identified a side product in a less polar fraction, the methyl ester of the starting acid, by NMR. He determined the mass of each of his fractions, and calculated an overall percent yield.

While there are some software programs that are making some inroads into the maintaining of a research notebook, these are primarily aimed at the industrial market. For some time to come, the hard copy research notebook should be an integral part of synthetic chemistry research in academe.

CHAPTER

8

CONDUCTING THE REACTION ITSELF

8.1 REAGENTS SUPPLIED AS DISPERSIONS

Some reagents that would normally be air-sensitive are rendered less so by being dispersed in mineral oil. Examples are LiH, KH, NaH, and Li. The mineral oil can cause problems in the workup (chromatography is required unless the product can be distilled), so it is a good idea to remove it before the reaction. This is done as follows: the necessary amount of dispersion (figure by the weight %, usually valid to within 5%) is placed in a dry flask under nitrogen containing a stir bar. The dispersion is covered with (suspended in) dry pentane or petroleum ether. The suspension is stirred for a minute and the stirrer is stopped. The reagent is allowed to settle, and the supernatant is removed by syringe or pipette. This procedure is repeated several times. Tipping the flask and letting the solvent flow off the reagent (which will clump and stick to the flask) helps remove all of the supernatant.

8.2 AZEOTROPIC DRYING

Some reagents, such as nucleosides, are so hygroscopic and hold onto water so tenaciously that it may be impossible to adequately dry them before they are placed in the reaction flask. In such cases in situ drying using an azeotrope may be effective. The reactant is dissolved in a solvent that forms a low-boiling azeotrope with water. Examples are collected in Table 4. While many chemists are most familiar with the benzene–water azeotrope, several other azeotropes use far less toxic organic solvents and contain a greater proportion of water. The solvent is evaporated on the rotary evaporator, provided that it is set up such that the vacuum can be released by

The Synthetic Organic Chemist's Companion, by Michael C. Pirrung
Copyright © 2007 John Wiley & Sons, Inc.

Table 4 Low-Boiling Water-Organic Azeotropes

Azeotrope	bp (°C)	wt % water
Water-pyridine	92.6	43.0
Water-toluene	85.0	20.2
Water-acetonitrile	76.5	16.3
Water-heptane	79.2	12.9
Water-benzene	69.4	8.9
Water-cyclohexane	69.8	8.5
Water-hexane	61.6	5.6

refilling the evaporator with an anhydrous atmosphere. If this capability is not available, the solvent can be evaporated on a vacuum line with an inert atmosphere refill capability. The solvent is stirred magnetically while vacuum is applied and the flask is held in a bath that provides the heat of vaporization. A cold trap condenses the evaporated solvent. To completely dry a reagent, evaporation from an azeotropic solvent several times may be required.

8.3 STOICHIOMETRY

An important decision to be made by the chemist concerns the exact amount of each reagent to use in a reaction. Consider the kinetic aldol condensation between ethyl isopropyl ketone and benzaldehyde (eq. 19). The enolate is generated by adding the ketone to lithium diisopropylamide (LDA), itself generated by treating diisopropylamine with a solution of *n*-butyllithium in hexanes. Benzaldehyde is added and the reaction is quenched with aqueous ammonium chloride. After aqueous workup, the *syn* aldol product is obtained. A balanced equation shows that each of the organic reagents is required in equimolar amounts, yet most chemists would not conduct the reaction that way. They would be more likely to use the ketone as the limiting reagent, 1.1 molar equivalents of *n*-butyllithium, 1.15 molar equivalents of diisopropylamine, and 1.0 molar equivalents of benzaldehyde. The rationale behind these choices is as follows:

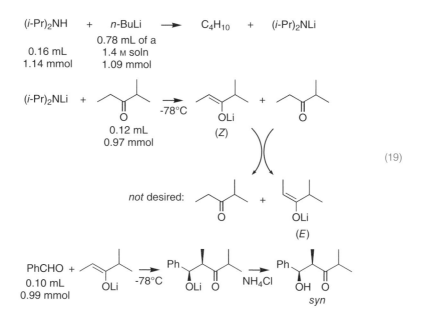

$$(19)$$

To obtain a single, kinetically defined enolate geometry, it is important to have LDA in excess during enolate formation. If any excess ketone is present it could serve as an acid to protonate the enolate. While this looks like a nonreaction, such proton transfers would in fact serve to equilibrate the enolate. That is, the enolate formation will no longer be under kinetic control, and the enolate stereochemistry will be defined by thermodynamic stability, which might not favor the single (Z)-enolate needed for the stereoselective aldol reaction. Since the ultimate base in this reaction is the n-butyl-lithium, it must be used in (a calculated) excess compared to the ketone. Use of a 10% excess allows for the actual titer of the n-butyl-lithium to be low. This might be due to inadvertent exposure of the solution to moisture since its last titration. It is also important to ensure that all of the n-butyllithium is consumed in generation of the LDA. If n-butyllithium remains after LDA formation, which might be due to an inaccurate (too high) titer or a fault in the

measurement or delivery of diisopropylamine, it can add nucleophili-
cally to the ketone or aldehyde. That is the reason for using an excess
of diisopropylamine compared to the *n*-butyllithium. Because diisopropylamine is basic (removable by an acid wash during extraction),
water soluble, and volatile, use of excess diisopropylamine creates
no difficulty in purifying the reaction product. Thus choices for the
amounts of reagents to use in this reaction trace directly to experimental uncertainties concerning their titer or the quantities actually
delivered. Finally, while these are the target stoichiometries, the
actual molar quantities used will be affected by the precision with
which volumes can be measured by syringe (see below). In equation
(19) are given the actual volumes of each reagent that would be used.
The slight variation of each molar quantity from the target is due to
the practical limitations on measuring and delivering the reagents.

The preceding example demonstrates how valuable the density of
a reagent can be to conducting a reaction, as it can simply be measured by syringe. If the density of a liquid is not available, it can be
determined as follows: Fill a 1 mL tuberculin syringe (with 0.01 mL
gradations) and needle (see the techniques following) with the liquid
and weigh it. Expel the liquid until the syringe reaches its stop, but
do not take any other action to expel the liquid in the needle. Because
a syringe is a TD (*to deliver*) volumetric device, it is made to deliver
a specified volume, so only by stopping at this stage can the volume
that was held in the syringe be known. Weigh the syringe again, and
determine the weight of the contained liquid by difference. This
method should be accurate to within 1%.

Reactions involving basic reagents and the generation of conjugate
bases from reactants are quite common. Consequently knowledge of
the acidities of a wide range of organic functional groups can be
essential to understanding and properly conducting synthetic reactions. It should also be noted that the acidities of compounds might
be quite different in aqueous media and in organic solvents, and that
it is impossible to directly determine the acidity of a compound that
is less acidic than water in an aqueous medium (as its conjugate base
simply deprotonates water). Acidities have been an active area of
research in physical organic chemistry for decades. Two compilations of functional group acidities, one relative to aqueous media and
one in DMSO, are provided in Appendixes 6 and 7. These are
representative and general data for the functional groups presented,

and can be modified by knowledge of the effects of specific structural variations on acidity (i.e., inductive, conjugation, and geometric/hybridization effects). The acidities of specific compounds are also available from several compilations. These compilations are frequently available only on the Web. While URLs for these sites will be provided, their continued access on the WWW is not ensured.

http://www.chem.wisc.edu/areas/reich/pkatable/index.htm

http://daecr1.harvard.edu/pKa/pka.html

The acid–base and chromophoric properties of such organic compounds can also be exploited to establish with certainty that reaction is complete. For example, triphenylmethide anions are intensely red in many ethereal solvents (an exception being lithium triphenylmethide in diethyl ether). Triphenylmethane can therefore be used as an indicator that base (it has been used with metal amides or the anion of DMSO) is present in excess in the deprotonation of any compound less acidic than itself (eq. 20). Compilations of the acidities of a variety of organic functional groups are presented in Appendixes 6 and 7. Another indicator is not dependent on its deprotonation. 2,2-Dipyridyl or 1,10-phenanthroline give a purple-red color in the presence of lithium dialkylamides in ethereal solvent. Therefore inclusion of these indicators in a reaction can demonstrate that a reagent (X–H) has been fully converted to its conjugate base (eq. 21). These indicators also show the presence of alkyl lithiums, as described in titrations (Section 2.4).

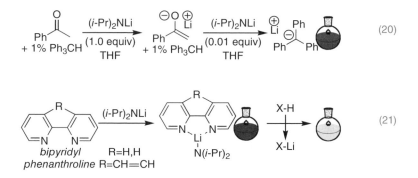

8.4 SYRINGE AND INERT ATMOSPHERE TECHNIQUES

There are three basic types of glass syringes. Tuberculin syringes with metal Luer lock fittings (Fig. 8.1) are preferable for working with dangerous or pyrophoric compounds. The Luer lock anchors the needle onto the syringe by screwing it in to the syringe. Multifit syringes (Fig. 8.2) usually have Luer lock fittings also, but they tend to leak between the barrel and the plunger. Tuberculin syringes with glass tips (Fig. 8.3) hold the needle on the syringe only by friction. This risks it coming off at an inopportune time. Glass syringes range in size from 100 to 0.5 mL. There is a greater tendency to leak around the plunger in the larger sizes (>20 ml), so they should be avoided when transferring hazardous compounds. During the transfer of such compounds (i.e., neat Me_3Al), a thin film of grease at the end of the plunger may provide a better seal and help to prevent clogging. For the transfer of larger volumes, cannulation may be preferable to the use of a large syringe (Fig. 8.4). A cannula is a needle with two points, or two needle points connected by a flexible length of Teflon tubing. It is used to transfer solution from one vessel

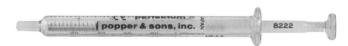

Figure 8.1 Glass syringe with Luer lock fitting. Copyright Popper & Sons, Inc.

Figure 8.2 Multifit syringe with Luer lock fitting. Copyright Popper & Sons, Inc.

Figure 8.3 Glass tuberculin syringe with ground glass tip. Copyright Popper & Sons, Inc.

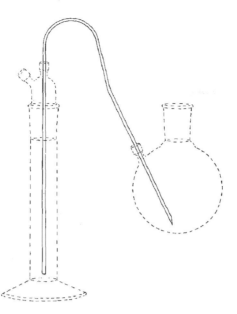

Figure 8.4 Transfer of a liquid using a cannula or double-ended needle.

directly into another. Plastic, disposable syringes with rubber plungers are also available. Testing of the integrity of these syringes to a given organic reagent or solvent is recommended prior to an actual transfer.

Typical stainless steel needles are 20 ga and come in lengths ranging from 3 to 45 cm. They are flexible and allow fast flow of non-viscous solvents. When clogging or slow flow is a problem (i.e., with organometallics or viscous solvents), a larger diameter needle (18 ga) is used. Care must be taken in using these needles because they are easily kinked, and very gentle bends are required. Needles larger than 18 gauge have thicker walls and so are not at all flexible.

8.5 GENERAL PROCEDURE FOR TRANSFER OF MATERIALS BY SYRINGE

Very light greasing of the needle is sometimes performed to allow it to puncture rather than tear the septum. The reagent bottle, flask, or reservoir may be attached to an inert gas source by a needle. This may be an inert gas line connected to a bubbler or a balloon filled with nitrogen. If the bottle is *not* attached to a gas source, the syringe should first be flushed (by repeated filling and expulsion) with inert gas and finally filled with gas to the desired volume of the transfer. This gas is then injected into the reagent bottle and the rest of the procedure is followed from step 2.

Step 1. Insert the needle through the septum but not into the liquid. Withdraw some gas into the syringe and remove the syringe from the bottle, grasping the needle as the syringe is pulled out. Expel the gas and repeat.

Step 2. Insert the needle below the surface of the liquid and withdraw more than the desired amount of liquid. Grip both the needle *and* the plunger. Gas pressure can push the plunger out of the syringe barrel, and needles not secured to the syringe by a Luer lock can fall off easily. There will almost always be a gas bubble in the syringe, so be sure to take steps to eliminate it. Invert the syringe, bending the needle into a U. The gas bubble should now be at the tip of the syringe. Expel gas and liquid until the syringe reads the desired volume. Pull the plunger back, withdrawing a bubble of inert gas into the syringe. This protects the liquid from exposure to air during transfer and prevents dripping. Grasp the needle and remove it from the bottle.

Step 3. Insert the needle through the septum of the receiver and hold the syringe with the tip up (a long needle may be needed to do this). Push the plunger in until it stops, injecting the small bubble of gas and the liquid that were in the syringe (dropwise if needed). The needle will remain filled with liquid. Pull back the plunger and remove some nitrogen into the syringe. Withdraw the needle and syringe from the receiver.

If a corrosive or chemically reactive reagent was transferred by a syringe, it is important to clean the syringe and needle by flushing them with a solvent immediately following the transfer. This prevents them from becoming clogged or frozen. For lithium reagents, hexanes can be used, which dilutes the reagent. The hexane washes should be carefully added to ethyl acetate and the syringe and needle flushed with the hexanes/ethyl acetate mixture several times. For Grignard reagents, ether should be used in place of hexanes, and the rest of the procedure followed as described. For other reagents, acetone can be used.

For the measurement and transfer of volumes smaller than 0.25 mL (250 µL), microliter syringes are used. These are usually available in 100, 50, and 10 µL sizes. Techniques are the same as above except that the needles are short and not flexible. The containers usually must be inverted to immerse the tip of the needle in the solvent and to expel all gas from the syringe. Alternatively, the plunger can be pumped several times with the needle in the liquid. These syringes sometimes have Teflon plungers, which must never be placed in the drying oven. They can be dried by placing disassembled syringes in a warm place.

Gastight syringes, which have Teflon plungers, are another major class. These are available in a wide range of sizes, from microliter to many milliliters. These are used for dangerous or air-sensitive reagents.

The pervasive use of syringes to add solutions of reagents and the convenience in measuring reagents based on molar concentration has induced chemists to use the same practice with a plethora of neat reagents, for example, diisopropylamine (used earlier in generating lithium diisopropylamide). The key to this practice is using the density of the compound to calculate the mass and therefore the moles (or, working backward from the moles to the volume). The density is frequently available in the chemical catalog from which the compound was purchased. The main potential snag with this approach relates to a reaction conducted at a temperature below the freezing point of the compound (also often available from the catalog). The neat reagent will turn immediately into a rock upon touching the cold solution, rather than dissolving as hoped.

8.6 ADDITION

If the order in which reagents are added has not been defined for a reaction, it is crucial to consider this question carefully. Often a reagent is already present in the reaction mixture and another reagent is added to it because it might be difficult to transfer, or simply because this is easier. A classical example of this situation is in Grignard reactions, where the organometallic reagent is first generated in an ether solution and then the electrophile is added. This is the so-called normal addition procedure. An "inverse" addition would involve addition of the Grignard to the electrophile. Transferring the Grignard naturally increases the risk of its exposure to the atmosphere and is not preferred. However, consider the addition of methyl Grignard to ethyl levulinate. In general, ketones are more reactive to nucleophilic addition than esters, so one would expect to be able to selectively add to the ketone to give the hydroxyester product (eq. 22). Yet, what would happen if ethyl levulinate were added to a methyl Grignard solution? When the first drop of ethyl levulinate hits the solution, undoubtedly the ketone would react with the Grignard first, but a large excess of Grignard reagent would still be present. It would certainly react with the ester as well, and the reaction outcome would not be the hoped-for selective addition. Now consider the reverse situation. What happens when the first drop of a methyl Grignard solution is added to the stoichiometric amount of ethyl levulinate? It adds to the ketone more quickly than to the ester, and all of the Grignard is consumed, so the ester remains intact. Only by adding the Grignard to the electrophile can the intended outcome be achieved.

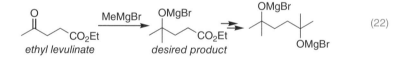

ethyl levulinate   desired product   (22)

There are at least two schools of thought on procedures for the addition of reagents. One is that solutions should be dropped directly into the reaction mixture so that there is no possibility of the reagent freezing, precipitating, or otherwise becoming heterogeneous so that it does not mix. Another is that solutions should be added along the wall of the flask so that they acquire the temperature of the reaction mixture overall before they mix with other reagents. This latter point applies particularly when the reagent is at room temperature and is being added to a cooled reaction mixture. It is also possible to add pre-cooled reagent solutions. On a large scale, this can be accomplished with a jacketed addition funnel, in which a coolant can surround the reagent solution. On a small scale, a double-ended needle (cannula) can be used. It is placed with one end in the solution to be transferred and the other end in the receiver. An inert gas at a higher pressure than in the receiver is used to push the solution through the needle. Whole reaction mixtures may be transferred in this way, even at low temperatures.

8.7 SPECIAL TECHNIQUES

8.7.1 Water Removal

Water is often the product of an organic reaction, for example in the elimination of an alcohol (eq. 23), a ketalization (eq. 24), or an esterification (eq. 25). The removal of this water will pull the equilibrium forward through Le Chatelier's Principle. A common way to remove it is via a solvent that forms a low-boiling azeotrope with water (heptane, toluene) and a Dean-Stark trap (Fig. 8.5). The azeotrope vapor is condensed into the trap. Because the aqueous and organic components are immiscible, layers form in the trap, and the

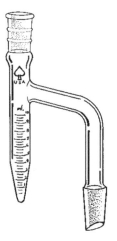

Figure 8.5 A Dean-Stark trap collects in the blind reservoir the water produced from a reaction. Volume graduations on this reservoir enable reaction progress to be followed by the amount of water that has been removed. In some traps the reservoir can be drained by a stopcock

lighter layer (the organic) eventually overflows back into the pot. The trap has volumetric markings that enable the volume of water to be measured (and in some cases a stopcock that enables the water to be removed). Reaction progress can be determined by the volume of water produced. In practice, it is often useful to wrap this whole assembly in aluminum foil to retain heat. It may otherwise be difficult to induce solvent vapors to reach the condenser. Another trick sometimes used is to add a small amount of water to the trap initially to ensure that layers are formed.

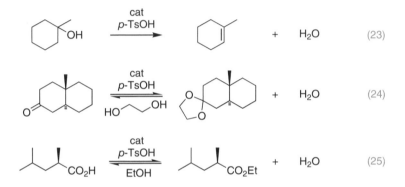

An alternative to the Dean-Stark trap uses a Soxhlet extractor (Fig. 8.6). Like the Dean-Stark trap, this apparatus benefits from being wrapped in aluminum foil to prevent radiative heat loss. The Soxhlet extractor repeatedly suspends a solid in a liquid in order to move a substance from one phase to the other. The solvent is heated to boiling, and the vapor moves up and is condensed into the upper reservoir. This reservoir contains the solid in a paper thimble, which is where phase transfer occurs. Once the reservoir is filled, the liquid overflows a siphon tube and the reservoir is drained into the flask below. A solute can be transferred from the solution to the solid, as intended for the removal of water from a reaction mixture using activated molecular sieves in the thimble. Another example uses a Soxhlet extractor to drive the exchange of carboxylic acids on a metal complex (eq. 26). The acetic acid forms an azeotrope with chlorobenzene and, upon reaching the thimble filled with sodium carbonate and sand, is irreversibly converted to sodium acetate. A substance can also be transferred from the solid to the solution, as when commercial copper iodide is extracted with tetrahydrofuran,

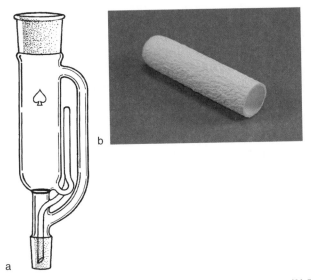

Figure 8.6 (*a*) The Soxhlet extractor, which is used with a thimble. (*b*) The thimble is made from rigid paper and is available in a range of sizes to fit different-sized extractors. The thimble contains the solid, preventing it from being carried by the liquid to the siphon tube or into the flask.

which removes CuI_2 from CuI because CuI_2 is much more soluble in tetrahydrofuran. This is an excellent method to purify CuI for the generation of organocuprates.

$$Rh_2(OAc)_4 + RCO_2H \rightleftharpoons Rh_2(O_2CR)_4 + AcOH \xrightarrow[Na_2CO_3]{PhCl} NaOAc \quad (26)$$

8.7.2 Reactions above Atmospheric Pressure

Some reactions must be conducted at pressures above atmospheric, for example, when the temperature required is far above the boiling point of the solvent or a reactant. While specialized equipment is available to conduct such reactions on a significant scale, they can be difficult to manage on our target scale of up to 25 mmol. One approach to this problem is the use of a sealed tube. A significant deficiency of such reactions is that they are impossible to follow. Heavy-walled Pyrex tubes (sometimes called Fischer-Porter tubes, often available from your friendly neighborhood glassblower) can tolerate pressures up to 20 atm. A glass rod is welded to the top of the tube using a flame. Reactants are added and the bottom of the tube is cooled in a dry ice or liquid nitrogen bath. The constriction in the tube is heated while the glass rod is pulled to draw down the constriction to closure and separation. Following the reaction, the bottom of the tube is again cooled (slowly and carefully! fracture is possible) in a dry ice or liquid nitrogen bath before the tube is opened, either by scoring and breakage or by a flame. Resealable pressure tubes (Fig. 8.7) are even easier to use but likely have lower pressure limits. Some laboratory microwave reactors (Section 5.1) are set up to handle and monitor pressures significantly above atmospheric, and these may be a much more convenient alternative to the sealed tube.

SAFETY NOTE

Further consultation with a glassblower on these methods may be worthwhile. Any pressurized system is hazardous and should be protected by a safety shield. Specific safety information concerning pressure tubes should be sought before any such reactions are attempted. Any defects (chips, scratches, bubbles) in these tubes can lead to explosions.

Figure 8.7 Resealable pressure tubes are more convenient to use than Fischer-Porter tubes but likely cannot tolerate pressures as great.

Another alternative are microwave digestion bombs made from Teflon (Fig. 8.8). These tolerate working temperatures up to 250°C and pressures up to 1200 psi and are sealed merely by turning the threaded cap. They accept reaction volumes of 23 to 45 mL and are made for use in microwave ovens.

8.7.3   Reagent Gases

The use of gases as reagents poses practical difficulties. For reactions requiring a gas only at atmospheric pressure, specialized equipment is not necessarily required, provided that the stoichiometry of the gas does not need to be controlled. Examples of such reagents include hydrogen, carbon monoxide, and carbon dioxide. The gas can simply be used in place of the usual inert nitrogen atmosphere. By the methods described in Section 4.4, the ambient atmosphere in the flask can be replaced by a pure reagent gas from a tank. Another alternative uses a balloon filled with reagent gas to replace the atmosphere of the reaction by repeated evacuation and filling. The balloon may be left connected to the flask during the reaction, providing a pressure slightly above atmospheric (sometimes called "positive pressure") and maintaining that pressure if a significant volume of gas is consumed. Recalling that a mole of an ideal gas is 22.4 L at

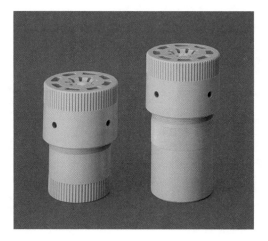

Figure 8.8 Teflon bombs for microwave heating under pressure. Photograph provided by Parr Instrument Company.

STP, even relatively small-scale reactions can consume significant volumes of gas. Particular care must always be exercised in concluding any reaction involving hydrogen, as the catalysts typically used for hydrogenation can also catalyze the reaction of hydrogen with oxygen in air, leading to a fire.

For reactions involving gases that require relatively low pressures, up to 5 atm, common pressure reactors include the Parr shaker (Fig. 8.9). These reactors are quite simple, with a heavy-walled glass pressure bottle sealed by a rubber stopper with a gas inlet. This whole assembly is held in a metal cage designed to hold the stopper in the bottle and contain the glass should the pressure bottle break. The cage is rocked by an oscillating arm driven by an electric motor, enabling liquid and gas to be mixed vigorously. The reactor has a gas reservoir, a vacuum inlet for evacuate-and-fill procedures, and a diaphragm valve to control the gas pressure. The main drawback of the Parr shaker is the volume of the smallest bottle (250 mL), which is inconsistent with the small-scale reactions of exploratory synthetic chemistry.

It may also be possible to generate a gas as needed for use in a reaction. For example, ozonolysis of alkenes is commonly performed in this way. Commercial ozone generators are available that use either air or oxygen and a corona discharge to produce ozone in a

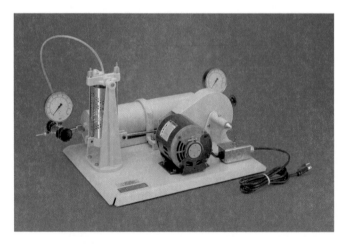

Figure 8.9 A Parr shaker for conducting hydrogenations under moderate pressure with shaking. Photograph provided by Parr Instrument Company.

gas stream. The toxicity of ozone to the human respiratory system demands that such ozone generators be used in an efficient fume hood. The ozone-containing gas stream is slowly bubbled into a solution of the reactant (in methanol, acetic acid, chloroform, hexanes, or ethyl acetate) to conduct ozonation. The rate of ozone production is defined by the input gas composition, its flow rate, and the electrical power, but is around 5 mmol/h with air and 0.5 mol/h with O_2. A calibration of the ozone generator can be developed by oxidizing a 5% aq. KI solution mixed with an equal volume of acetic acid and titrating the so-produced I_2 with sodium thiosulphate solution to a starch endpoint. This gives the experimenter a good estimate of how many mmol of ozone can be generated per minute at specific instrument settings and therefore how long a particular ozonolysis reaction should require.

When only one ozone-reactive function is present in the molecule, ozonation to exhaustion is simple to conduct. This approach exploits the fact that solutions of ozone in organic solvents are blue. So long as the ozone is consumed by reaction, the solution will remain colorless. When all of the reactant is consumed, the ozone dissolves, demonstrating the end point with its blue color. Adding a small complication to all this is the fact that ozonolysis of alkenes is generally conducted at −78°C (because ozone reacts explosively with alkenes at room temperature). Therefore these observations must be made through a dry ice cooling bath. Once the end point is reached, the ozone stream can be replaced with a stream of nitrogen and the excess ozone can be swept out of the solution. Workup of the ozonolysis reaction requires reduction of the ozonide at the low reaction temperature, since the ozonide is also a peroxide and such compounds can be dangerously explosive (see Section 6.2).

Oxygen gas can also be used as a reagent, either in radical reactions as its ground-state triplet or in cycloaddition reactions as its excited singlet state. Simply exposing a reaction mixture to air typically does not promote an efficient reaction with O_2. Air must be bubbled through the solution or an oxygen-rich atmosphere must be used. If air is used, moisture and carbon dioxide may need to be removed with a KOH drying tower if the reaction would be sensitive to these species (e.g., oxygenation of an enolate). The generation of singlet oxygen in an oxygenated solution requires light (typically visible light from a sunlamp) and a triplet sensitizer (like tetraphenylporphyrin or Rose Bengal). The generation of singlet oxygen may be an unwanted side process during photochemical reactions, providing another circumstance in which degassing (deoxygenation) of the solvent is important. Pure O_2 gas should be used very carefully because it can promote combustion reactions that are not a concern at the lower partial pressure of O_2 in air. Also, since the products of most reactions involving O_2 are peroxides, which as a class are reactive and often explosive, such reactions must be conducted with care and appropriate precautions, such as a safety shield.

8.7.4   Ultrasonication

The use of ultrasonic energy may be recommended for some reactions. The ways in which ultrasound can potentially influence reactions have been extensively discussed but will not be belabored here. Reactions for which there is a clear benefit include those involving solid–liquid interfaces or activation of metal surfaces, such the formation of Grignard or organozinc reagents. The effects of ultrasound may include removing oxide coatings from the metal surface, exposing reactive sites of the clean underlying metal, and facilitating reactant transport to and from the solid phase. In fact ultrasonication is believed to be the most effective method of mixing known. The ability of ultrasound to affect a reaction is clearly dependent on the strength of the sonic energy source. Sophisticated ultrasonication instruments may be available in specialized laboratories, but many labs have ultrasonic cleaning baths. The effectiveness that such a bath should have on a chemical reaction can be easily tested. A sheet of aluminum foil is placed in the bath for 30 seconds. If it emerges pockmarked with holes, the bath is strong enough to affect a reaction.

8.8   QUENCHING

This term has acquired a variety of meanings and uses, some proper, some not. In principle, it refers to adding a substance that deactivates reagents present in the reaction mixture that might prevent product isolation. For example, a Grignard addition to a ketone initially produces a magnesium alkoxide. In order to isolate the desired alcohol, acid is added to the reaction mixture, which protonates any remaining Grignard as well as the alkoxide (eq. 27). This is appropriately called an acid quench. Loose uses of the term include any solution added following the end of the desired reaction in preparation for the workup.

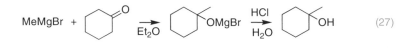

$$\qquad\qquad\qquad\qquad\qquad\qquad\qquad\qquad\qquad\qquad (27)$$

8.9 SPECIALIZED REAGENTS

8.9.1 Diazomethane

Diazomethane is attractive as a methylating agent for carboxylic acids and phenols because it reacts quickly and highly efficiently with the production of only N_2 as a by-product (Black, 1983). Its natural yellow color is discharged as it reacts, providing automatic indication of reaction progress. However, because diazomethane is highly toxic, it should be generated and used only in a well-functioning fume hood. Because it explodes on contact with groundglass of any type (joints, stoppers, syringes, stopcocks), it should be handled behind a safety shield, and other personal protective equipment should be used. Because it has a boiling point of −23°C, it is usually handled in the ethereal solutions in which it is generated. Because it explodes on contact with $CaSO_4$, its solutions or vapors must never be dried with DRIERITE. Despite all of these hazards it can be worked with safely, provided that appropriate precautions are observed.

Two main methods are used to prepare diazomethane. One uses commercially available apparatus specifically designed for its preparation and distillation while entrained with ether. The resulting ether solution is typically of 0.3 to 0.4 M concentration and is diazomethane in its purest form. Such apparatus have specialized joints without ground glass and come in a range of sizes for generating diazomethane on scales of around 1, 50, or 300 mmol. The other method uses conventional glassware. Both methods use hydroxide to generate the

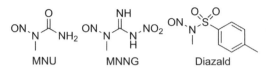

Figure 8.10 Reagents used as precursors to diazomethane.

diazomethane from nitrosamide precursors. The more formal method involves adding *N*-methyl-*N*-nitroso-toluenesulfonamide (Fig. 8.10), also known as Diazald, to KOH. The manufacturer's instructions for the use of these apparatus should be followed explicitly.

The home brew method can be found in its original form in *Organic Syntheses, Collective* Volume 2 (using *N*-methyl-*N*-nitroso-urea as the precursor). This method uses a two-phase system of 50% aq. KOH and diethyl ether in an Erlenmeyer flask cooled in an ice-water bath and stirred magnetically. The precursor recommended today, because it is safer to store and handle, is the crystalline solid *N*-methyl-*N*-nitroso-nitro-guanidine (MNNG). However, MNNG is still considered toxic, a severe irritant, a carcinogen, and a mutagen, and is typically used for generation of diazomethane quantities of 1 mmol. MNNG is slowly added to the two-phase system portion-wise. Sufficient precursor must be used to allow for materials transfer losses that are inevitable in the incomplete separation procedures described following. A yellow color will develop in the ether phase as the diazomethane is generated. After all of the precursor has been added, the solutions may be stirred for 10 minutes or so to allow the reaction to complete. The upper ether layer is decanted into a clean flask held in an ice-water bath. DO NOT use a separatory funnel with a ground-glass stopcock to separate the aqueous solution from the ether phase. Another portion of ether is added to the reaction flask, and it is stirred at ice-water bath temperature to extract remaining diazomethane. This ether layer is also decanted into the clean flask in the ice-water bath. This process may be repeated. The combined ether phases are likely to contain some dissolved water, which may be removed by adding KOH pellets and allowing the solution to stand in an ice-water bath for 0.5 to 3 hours. The resulting yellow ethereal solution of diazomethane is ready for use. This procedure can be conducted on up to a 60 mmol scale.

8.9.2 Lithium Aluminum Hydride

LiAlH$_4$ can be readily handled as the powder in air without loss of active hydride, but is an excellent reducing agent in ethereal solution. Chemists typically think of the reagent as an ionic species (Li$\oplus$ and the complex ion AlH$_4\ominus$), which suggests that its solubility would be greatest in more polar ethereal solvents like tetrahydrofuran (THF). However, the solubility of lithium aluminum hydride in ethereal solvents has been determined, and it is twice as soluble in diethyl ether as in THF.

A common transformation using LiAlH$_4$ is the reduction of carboxylic acids to primary alcohols. Because this reaction necessarily proceeds by initial deprotonation of the acid by a hydride from LiAlH$_4$ to give a carboxylate, there may be an inclination among chemists to use tetrahydrofuran as the solvent for such an ionic species. It might also be thought that the carboxylate would be quite resistant to reduction due to its negative charge and that the reaction therefore requires heat, with the higher boiling point of THF providing an advantage. Yet such reductions often proceed as well or better in ether than in THF. A potential reason for this observation is the greater solubility of the reagent in ether.

LiAlH$_4$ ethereal solutions are gray in color, similar to the powder. If they are filtered (under a nitrogen atmosphere, of course, using a Schlenk-ware coarse fritted filter, Fig. 8.11), the filtrate is often much more transparent, and a fine gray residue is left behind on the filter. The residue often exhibits some active hydride, so it must be quenched with care, but the amount of active hydride in the filtrate is typically not appreciably diminished. This filtered ethereal LiAlH$_4$ solution may afford much cleaner reduction reactions. Lore concerning these observations is that Lewis acidic impurities that are insoluble in ether are removed by the filtration process. While evidence about the basis of the effect may be debated, such effects are clearly real in some instances.

8.9.3 Hydrogen Peroxide

This reagent is one of the least expensive and most convenient oxidants known. It is used in a variety of transformations, including the epoxidation of α, β-unsaturated carbonyl compounds and the

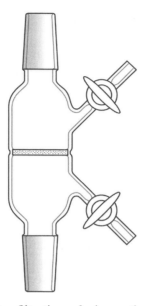

Figure 8.11 Apparatus for filtration of air-reactive solutions under an inert atmosphere.

preparation of peracids. H_2O_2 is typically available as an aqueous solution in several strengths, including 3%, 30%, and 90%. The first two should be handled with respect, like any chemical reagent, but do not pose undue hazards. Like other peroxides care must be taken in exposing H_2O_2 to even mildly reducing conditions, which can initiate radical reactions. Chemists should be more concerned about concentrated H_2O_2 solutions (Shanley and Greenspan, 1947), which are commonly used for the in situ generation of peracids like trifluoroperoxyacetic acid from the anhydride. Contact of hydrogen peroxide of more than 65% concentration with any combustible material (cotton, wood) leads to fire. Pure 90% H_2O_2 does not explode, but matters are different in the presence of a fuel or catalyst. Metal-containing reagents may react violently with H_2O_2. The exothermic decomposition of 90% H_2O_2 yields 5000× its volume in O_2, creating the risk of pressure buildup. The combination of glycerol and 90% H_2O_2 is comparable to nitroglycerin as an explosive, except that the former is 7× more impact-sensitive. Iron, brass, or copper fittings or mercury thermometers should not be used with high concentration H_2O_2 because the metals can promote its decomposition.

8.10 REACTION TIME VERSUS PURIFICATION TIME

This chapter closes with a broad admonition concerning synthetic reactions. Attention to detail in the execution of the reaction up to this stage will pay enormous dividends in the chemist's time. Conduct of the reaction itself is often a small fraction of the time spent in isolating and purifying the product. Any step the chemist can take to make a reaction more selective and/or eliminate a by-product will be well worth it.

CHAPTER

9

FOLLOWING THE REACTION

It is essential to follow the progress of the reaction whenever this is possible. Each time point in a reaction at which the chemist has some indication of what is happening is like a separate experimental result. The only other way to get this type of information would be to run replicate reactions for those periods of time, work them up, and determine the composition of the reaction mixture through analytical/ spectroscopic techniques. Obviously, following one reaction is far easier than conducting and analyzing many. It is a good practice to save a small sample of starting materials as a reference and to obtain analytical/spectroscopic information on all reactants before the reaction is begun. Reactions that are new to the experimenter should be monitored at 5 minutes, 15 minutes, 30 minutes, 1 hour, 2 hours, 4 hours, 8 hours, and so forth. It is easy in a busy research laboratory to forget to take a data point to track a reaction. Chemists could take a lesson from the biology laboratory here and use small, inexpensive electronic timers that can be programmed for multiple time points. Several analytical methods are available for following reactions. The value of simple observations should also not be overlooked. Color or phase changes (precipitation or dissolution) may signal that a reaction has occurred. Temperature increases may occur because of exothermic reactions, a reason to include a thermometer that extends into the solution inside most reaction setups.

9.1 THIN LAYER CHROMATOGRAPHY (TLC)

Commercial TLC plates are generally large (20×20 cm) and come with several choices of backings, including glass and aluminum sheet. Many chemists prefer glass. While some glass TLC plates are pre-scored for breaking into particular sizes, this prevents the chemist from preparing plates customized to a particular use or from

The Synthetic Organic Chemist's Companion, by Michael C. Pirrung
Copyright © 2007 John Wiley & Sons, Inc.

economizing by making smaller plates. It is better to obtain unscored plates and cut them as desired. TLC plates come with a wide variety of adsorbents. The most common are silica gel 60 plates, which are optionally available with a fluorescent indicator that facilitates ultraviolet (UV) detection. Also available are alumina and reverse phase (see below for discussion of this concept) adsorbents.

9.1.1   Cutting Glass TLC Plates

1. Use a guide rack or straightedge to ensure that parallel lines and appropriate spacing are maintained.
2. Two types of glass cutters are available—the wheel type and the diamond type. The wheel type must be lubricated with a drop of mineral oil.
3. Clean the glass surface of the plate with a tissue.
4. Using the guide, score the glass with the cutter at desired intervals.
5. Make all scores by moving the cutter in the same direction, pressing hard with a wheel cutter, firm with a diamond cutter.
6. Turn the plate 90° and repeat.
7. If a wheel cutter was used, clean the oil off plate with a tissue moistened with pentane.
8. Break the scored plate into individual TLC plates by placing thumbs on either side of each scratch, protecting the adsorbents side of the plate with a paper towel, and bending up.

9.1.2   Spotting TLC Plates

A pencil and straightedge are used to mark a line on the sorbent higher than the solvent level in the TLC chamber. Samples are then spotted along that line. TLC spotters are easily prepared by drawing out capillary tubes with two open ends over a flame. A little practice and experience should guide how long and narrow to draw the capillary. Almost all samples for TLC should be spotted from solutions of a milligram or so in a relatively volatile solvent. Spotting neat liquids will overload the plate. NMR samples are quite good for TLC spotting. Reaction mixtures are readily spotted simply with whatever is drawn into a capillary spotter. If the reaction is being conducted in a nonvolatile solvent (HMPA, DMF, DMSO), a large baseline spot will remain because these very polar solvents are poorly eluted. The mobility of the reactants and products might also be affected because of dissolution of these solvents in the eluent as it moves past the baseline spot. However, it may still be possible to follow by TLC reactions in these solvents, especially if they are at least mostly removed following spotting. This can be done by heating with a heat gun and/or pumping off inside a filtering bell jar (Fig. 9.1). The TLC plate is held upright in a small beaker that is placed in the bell.

It is good practice to save a small sample of each synthetic intermediate in vials for future TLC comparisons. Authentic samples of the nonvolatile reactants or other known compounds should be spotted on the plate with the reaction mixture to provide more information concerning the identity of a particular spot. This spotting can be done in parallel tracks or even on the same track (called co-spotting). The co-spot aids the differentiation of reactants and products when they have similar R_fs, and helps diagnose changes in the appearance of the TLC of a reactant when spotted from the reaction mixture. It occasionally happens that a starting material appears different on TLC when spotted from the reaction mixture rather than from a pure sample. This can mislead the chemist into thinking that the starting material has been consumed when it has not.

Figure 9.1 A filtering bell jar can be used to apply vacuum to a spotted TLC plate to remove poorly volatile reaction solvents. Photograph provided by Kimble/Kontes.

9.1.3 Eluting TLC Plates

A wide variety of TLC chambers are available, and their design is not crucial, provided that they have a flat bottom and are narrow enough that the TLC plate cannot fall flat. Their primary function is simply to provide a thin layer of eluent solution to be drawn up the plate and an atmosphere saturated with eluent vapor. For the latter purpose, a filter paper is usually adhered to the wall of the chamber as a wick. After the solvent front has moved almost to the top of the absorbent, the plate is removed and the solvent front is marked with a pencil for the determination of the R_fs. For standard analytical TLC, the polarity of the eluting solvent should be adjusted such that the R_f of the compound of interest is around 0.5. This permits products that are either more or less polar to be easily observed. If the starting material spot is at the extreme bottom or top of the plate, products cannot be observed unless their polarity changes in the right direction. Of course, this rule may be violated if the polarity of the products relative to the starting material is already known.

TLC usually utilizes mixtures of a more polar and a less polar solvent so that eluent polarity can be readily adjusted by changes in the volume proportion of the two solvents. Some solvent pairs commonly used for chromatography in synthetic chemistry are summarized in Table 5. A problem common in the TLC of quite polar compounds is "streaking" (Fig. 9.2). It can be difficult to distinguish one compound that streaks from two compounds that have close R_fs. A common remedy for streaking is inclusion in the eluting solvent of low levels of a component with the same chemical character as

Table 5 Commonly Used Chromatography Solvent Mixtures

Hexane/ethyl ether	Standard separations; nonpolar compounds
Hexane/ethyl acetate	Standard separations; more polar compounds
Hexane/ tetrahydrofuran	Standard separations; yet more polar compounds
Dichloromethane/ methanol	Very polar compounds
Chloroform/MeOH/ 1% aq. NH_3	For amines

Figure 9.2 Polar compounds can give unsymmetrical spots, called streaks or tails, on TLC.

the compound being separated. That is, streaking of amines is reduced by ammonia in the eluent, and streaking of carboxylic acids is reduced by formic or acetic acid in the eluent.

Normal phase TLC involves a hydrophilic stationary phase (silica gel, alumina) such that more polar solvents have greater eluting power. A series expressing the relative eluting power of a selection of solvents has been developed and is summarized graphically (Fig. 9.3). Reverse-phase TLC involves a hydrophobic stationary phase (silica covalently bonded to C4, C8 (Fig. 9.4), or C16 hydrocarbon chains) such that less polar solvents have greater eluting power. Mobile phases commonly used in reverse-phase TLC (or HPLC) include water/methanol and water/acetonitrile mixtures. The relationship between the relative eluting power of the two mixtures is not linear, but a nomograph (Fig. 9.5) can be used to relate the mobility of a compound in two different solvent systems.

One problem that sometimes arises in TLC (and other absorption chromatography) is that the compound may be unstable or otherwise chemically modified during the elution. A simple way to test for this possibility is two-dimensional analysis. The sample is spotted in one corner of a square TLC plate. It is eluted in one direction, dried, turned 90°, and eluted again. It is then visualized. If the compounds have survived the TLC without modification, all spots should be found along the diagonal (Fig. 9.6). Off-diagonal spots indicate that modification has occurred during the elution.

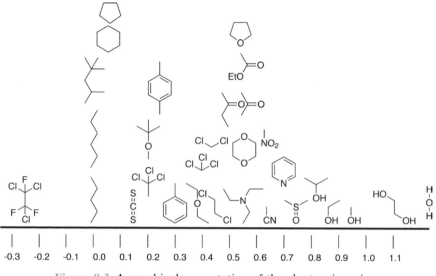

Figure 9.3 A graphical presentation of the eluotropic series.

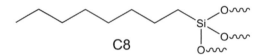

C8

Figure 9.4 Reverse-phase silica gel has long-chain hydrocarbons chemically bonded to the silica backing (through a *tris*-siloxane group and the wavy bonds shown).

If compounds prove difficult to separate on TLC, multiple elutions may help. A solvent with about a third of the polarity of that in which $R_f = 0.5$ is chosen. The plate is eluted and then dried before the next elution; this process may be repeated up to four times.

9.1.4 Visualizing TLC Plates

The first analysis of an eluted and dried TLC plate should be with a long-wavelength UV light ("black" light; Fig. 9.7). This method is nondestructive, whereas analyses with dips, sprays, or stains are irreversible. Compounds with UV chromophores that are fluorescent (primarily aromatic rings) may show up as vivid fluorescent

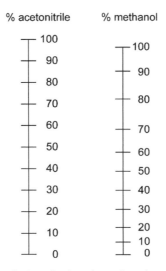

Figure 9.5 A nomograph relating the fraction of methanol or acetonitrile in water that provides equivalent mobility on a reverse phase sorbent.

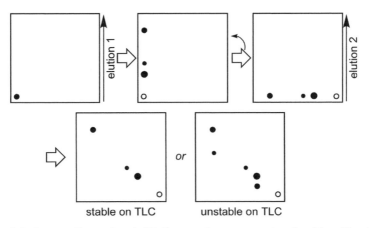

Figure 9.6 A two-dimensional TLC experiment permits the identification of compounds that are unstable on silica gel.

spots under UV light, but this is not very common. If the TLC plates include a fluorescent indicator, compounds with UV chromophores that are not fluorescent (i.e., conjugated enones, dienes, or aromatic rings) can quench the fluorescence and show up as dark spots on a green background. Manufacturers typically use descriptors like

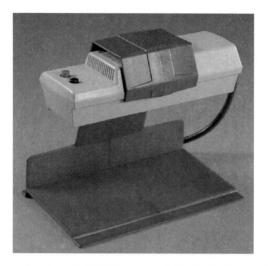

Figure 9.7 A long-wavelength UV lamp used for initial visualization of TLC plates. Photo provided by UVP Inc.

"silica gel F_{254}" to indicate TLC plates including a fluorescent substance with a 254 nm absorption.

A chamber containing iodine crystals will stain many organic compounds, and because of iodine's volatility, staining may be reversed by removing the plate from the chamber. Some indication of functional groups present may be obtained from iodine stains. Alkenes stain strongly, which is readily understood based on their reaction with halogens, and other unsaturated compounds are usually reliably stained. Saturated hydrocarbons, esters, ethers, and nitriles usually do not stain well, and many halogen compounds give a "negative" spot: a white zone on the slightly stained silica.

Other visualization methods use chemical sprays or dips. A wide range of stains are collected in Table 6. Recipes for the preparation of the various TLC stains are provided in Appendix 3. Dipping into a beaker (stored in the hood covered with a watch glass) filled with solution avoids toxic and corrosive chemical vapors in the lab from sprays. Using a forceps, the TLC plate is dipped into the solution up to the solvent front and removed. The back of the plate is cleaned with a tissue, and the plate is heated on a hotplate in the hood or with a heat gun. Spots usually develop in less than two minutes.

Table 6 Stains for the Visualization of Spots on TLC Plates

Dip or Stain	Comments
Phosphomolybdic acid (PMA)	General organic compound stain, plates must be heated to develop spots, does not distinguish different functional groups, spots are usually dark green on a light green background, alcohols typically give strong spots
Vanillin	General stain, amine, hydroxyl, and carbonyl compounds show strongly, spot colors are compound-dependent, smells like cookies
Dinitrophenylhydrazine (DNP)	General stain, identifies aldehydes and ketones by forming the corresponding hydrazones, which are usually yellow to orange
Cerium molybdate (Hanessian's stain)	General stain, usually requires vigorous heating, spots are often dark blue (though colors may vary) on a light blue or light green background
p-Anisaldehyde stain **A**	General stain, uses mild heating, sensitive to most functional groups, especially those that are nucleophilic, can be insensitive to alkenes, alkynes, and aromatic compounds, spot colors are variable on a light pink background
p-Anisaldehyde stain **B**	More specialized stain than **A**, uses mild heating, developed for terpenes
KMnO$_4$	Stain for functional groups sensitive to oxidation, alkenes/alkynes stain rapidly at room temperature as bright yellow spots on a purple background, other oxidizable functional groups (alcohols, amines, sulfides, and mercaptans) stain with gentle heating, their spots are usually yellow or light brown on a light purple or pink background, spots should be circled following visualization as the background becomes light brown with time
Ninhydrin (commercial spray reagent available)	Selective for primary, secondary, and tertiary amines, amino sugars, classical detection for amino acids, take care not to get it on skin
Bromocresol green	Selective for functional groups such as carboxylic acids with $pK_a < 5$, bright yellow spots on either a dark or light blue background, heating is not typically required

Table 6 (*Continued*)

Dip or Stain	Comments
Ferric chloride	Selective for phenols
Ceric ammonium molybdate	Strong stain for hydroxy compounds
Ceric ammonium sulfate	Specifically developed for vinca alkaloids
Cerium sulfate	General stain, particularly effective for alkaloids
Ehrlich's reagent	Selective for indoles, amines, and alkaloids
Dragendorff-Munier stain	Selective for amines, and will stain amines of low reactivity that other stains do not

Caution must be exercised in interpreting the results of any TLC visualization, especially regarding the proportions of compounds in a mixture. Compounds that stain darkly or as large spots with a particular stain may be minor components and compounds that stain lightly or not at all may be major components. Indeed the whole purpose of some stains is to visualize some compounds selectively, most often based on the reaction of specific functional groups with reagents in the stain. The size of a spot on a TLC plate may have no direct relationship to the proportion of that component in the sample. The chemist may be seriously misled by attempting to "integrate" a TLC plate.

When applying any visualization method to a TLC plate, it is a good idea to circle the observed spots with a pencil so that the exact position and size of the spots is preserved for following visualization steps. This information can aid in identifying compounds that are visualized by two different methods and help distinguish spots of compounds with close R_fs but different staining properties. Circling the spots also deals with the common observation that spots fade over time. Depending on the particular stain used, compounds bearing different functional groups may or may not be distinguished based upon the spot colors. Nevertheless, it may be worthwhile to note the colors of spots of particular compounds so that when similar chemistry is performed on related compounds, correlations of spot colors with functional groups may be discerned.

When following a reaction by TLC, it is generally assumed that any anionic or metalated intermediates in the reaction mixture will be protonated upon spotting onto the TLC plate (eq. 28) and will migrate as the conjugate acids. In some cases, where reaction components interfere with TLC analysis, following a reaction may involve removing small aliquots and subjecting them to mini workups to obtain a solution that can be analyzed.

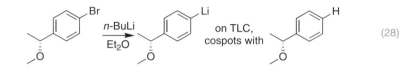

(28)

9.2 GAS CHROMATOGRAPHY (GC)

The capillary column gas chromatograph with flame ionization detector (FID) is an excellent analytical tool, capable of analyzing products with sensitivity far greater than usually needed to follow organic reactions (i.e., ng). As the name implies, the column is a fused silica capillary 30 to 100 m in length whose inside surface is coated with the stationary phase. The primary basis of separation is the boiling point. Small samples of about 1 µL of solution are injected with a special GC syringe. Demonstrating the sensitivity of the instrument, much of the injected sample is split off to waste to prevent overloading the column. Crude reaction mixtures might be injected, but if salts or other nonvolatile compounds that can foul the injector are present, they are usually removed by filtration through a polar sorbent like silica or alumina. Typically the column is temperature programmed, starting at a fairly low temperature (ca. 100–150°C) and ramped up to high temperature to enable the analysis of a wide range of compounds. The mobile phase is a dry carrier gas, He. The FID uses H_2 and air to generate a flame that ionizes the column effluent, and it is this ionization that is detected. This detector thus responds to the ability of a compound to burn. One expects compounds that are isomers to have quite similar properties in this regard, and therefore their peak sizes should be directly proportional to their representation in the sample. Peaks for two compounds with different abilities to burn will not be directly reflected in their peak magnitudes (specifically, integrated area). If one wants to accurately measure relative amounts of two dissimilar compounds (e.g., for a kinetic study), it is necessary to determine the relative response of the detector to each compound. This is not often done for routine organic reaction monitoring.

9.3   HIGH-PRESSURE LIQUID CHROMATOGRAPHY (HPLC)

Analytical HPLC is a more broadly applicable method to follow reactions than GC because some compounds are simply not thermally stable or volatile enough for GC. Two major types of HPLC columns are used. Normal-phase HPLC is based on a polar sorbent, typically silica gel. This method is like conventional absorption chromatography, with the advantage that the resolving power is much greater because the sorbent has much finer mesh, requiring high pressure to push the eluent through the column. The eluent can be a single composition (isocratic) or involve an increase in the more polar component as the run proceeds (gradient). Like temperature programming in GC, elution gradients permit compounds of a wide variety of properties to be analyzed in a single HPLC run and ensures that all compounds loaded onto the column are eluted. Computer-controlled pumps mix the solvents in the correct proportion for the gradient. Reverse phase HPLC is based on a nonpolar sorbent generated by derivatizing silica gel with hydrocarbon groups (Fig. 9.4). The eluting solvent is typically a mixture of water and either methanol or acetonitrile, and the gradient involves an increase in the organic component.

Types of detection for HPLC include UV (both single wavelength and diode array detectors, which can collect a full UV spectrum on the eluent at any point in time) and refractive index. The strength of the detector signal for a specific compound will depend on the difference between its UV absorption or refractive index and that of the solvent. Refractive index is a property that saw quite a bit of use in organic chemistry in the past but is hardly used today. Compounds with minimal functionality tend to have lower refractive index, while more functionalized and unsaturated compounds tend to have higher refractive index. Typical chromatographic solvents used with refractive index detectors are diethyl ether and hexanes. HPLC-MS instruments have recently become available at a price that makes them practical for individual departments or even laboratories to acquire.

LC-MS has become much more widely used as a powerful method to identify reaction products.

9.4 NMR SPECTROSCOPY

It is often possible to conduct a reaction on a relatively small scale in a solvent commonly used for NMR, such as d_6-benzene, d_3-chloroform, d_6-acetone, d_6-DMSO, d_3-acetonitrile, d_3-methanol, or D_2O, and observe reaction progress spectroscopically. Techniques that suppress solvent protons may also be used with nondeuterated solvents (Hoye et al., 2004). For proton NMR, integrations of the peaks in the spectrum should be accurate to within about 10% of the actual compound ratios in the sample. Reactions that involve paramagnetic reagents or impurities or include suspended solids may be difficult to analyze by NMR because they give poor spectra. Particularly for the chemistry of air-sensitive transition metal complexes, where none of the chromatographic methods for following reactions is applicable, NMR is a prime means of reaction monitoring. Special NMR tubes that can be sealed with a flame under an inert atmosphere are often used for following transition metal transformations.

CHAPTER

10

WORKING UP REACTIONS

The first consideration in working up a reaction is to remove any stopcock grease that was used on the joints. A tissue moistened with a solvent such as hexanes or dichloromethane does this well.

A significant issue in planning the workup is whether solvent partitioning should be used. The main purpose of aqueous/organic partitioning is the removal of inorganic salts produced in a reaction, along with any water-soluble solvents, reagents or by-products. If none of these is present, there is little point in solvent partitioning, and it may provide an opportunity for loss of water-soluble products. Evaporation of the solvent and direct purification has much to recommend it. In some cases compounds may not be able to survive aqueous workups, or one may simply choose not to do one. If aqueous/organic partitioning is needed and the reaction was conducted in a polar organic solvent, such as small alcohols or tetrahydrofuran, a worst-case scenario is that phase separation does not occur during partitioning because the polar organic is miscible with both water and the extraction solvent. Even if two phases are formed, often the presence of the polar organic solvent enables some product to dissolve in the aqueous phase and thereby be lost. In such cases it may be advisable to evaporate the polar solvent before partitioning the residue between water and a volatile organic solvent nicely immiscible with it.

Having completed a reaction that included a magnetic stirring bar, it can be a challenge to pour the reaction mixture into a separatory funnel without also having the stir bar end up in the funnel. Special retrievers (magnets on inert plastic cords) can be used to remove the stir bar from the reaction before transferring it to the funnel. Some dexterous chemists merely hold a powerful horseshoe magnet near the neck of the flask as they pour out the reaction mixture, catching the stir bar as it comes by.

The Synthetic Organic Chemist's Companion, by Michael C. Pirrung
Copyright © 2007 John Wiley & Sons, Inc.

For reactions involving highly polar, poorly volatile solvents like ethylene glycol, DMSO, DMF, and HMPA, getting rid of the solvent can be quite a challenge. It may be possible to remove some of them by evaporation before solvent partitioning, though high vacuum (and even warming) may be required, and not all products can tolerate this treatment. One good strategy is to partition the reaction mixture between an aqueous solution and a hydrocarbon solvent such that the polar reaction solvent prefers the aqueous phase. However, this ploy only works when the desired reaction product partitions predominantly into the hydrocarbon. This strategy can also be applied to reactions in acetic acid, an approach far superior to using large quantities of hydroxide (or bicarbonate, with voluminous CO_2 evolution!) to extract the acetic acid. Of course, a base wash should still be used to ensure that all of the acetic acid has been removed from the organic phase.

Washing polar/acidic reagents away as an initial step in the workup is a profitable strategy. The condensation of lactic acid with acetone to give a dioxolanone is promoted by stoichiometric boron trifluoride etherate (eq. 29). If this reaction were simply quenched with water, large quantities of fluoboric and hydrofluoric acids would be generated that would require large quantities of base to neutralize. Instead, the reaction mixture is diluted with ether and washed with aqueous sodium acetate. Some of the acid simply partitions to the aqueous phase, and some reacts with the acetate to form acetic acid, which partitions to the aqueous phase. The quantity and strength of the acid that must be washed out of the organic phase with base is thereby much reduced.

$$(29)$$

10.1   SOLVENT EXTRACTION

The goal is to ensure that all of the desired product finds its way to the organic phase; as many and as much as possible of the by-products should go to the aqueous phase. Minimizing the volume of the aqueous phase, consistent with dissolving all of the salts, will maximize the organic product going to the organic phase. Extraction with solvents heavier than water makes the process operationally easier than the converse, but presents the problem that most such solvents are toxic chlorocarbons. Dichloromethane is the least noxious of the bunch. Other common extraction solvents are diethyl ether, hydrocarbons, and ethyl acetate. With molecules particularly difficult to extract from the aqueous phase, mixed solvents may prove helpful. For example, a mixture of 10–15% isopropanol in a chlorocarbon is one of the most polar organic solutions that is immiscible with water. It is quite effective in extracting polar organics such as phosphonates or heterocycles (Fig. 10.1). To ensure easy phase separation when using such solvent mixtures, it is essential to consider densities such that one phase is significantly more or less dense than the other. A qualitative measure of all of the factors contributing to solvent miscibility and therefore utility in phase partitioning (including chromatographic methods) has been developed as the mixotropic solvent series (Appendix 4). The farther two solvents are apart in this series, the less miscible they are.

The aqueous phase typically is extracted a few times with equal volumes of the organic solvent and the combined organic phases are washed. The wash solutions vary with the reaction: $NaHCO_3$ solution and 10% NaOH solution will remove acidic compounds, depending on their pK_a, while 5% HCl is used to remove bases. $NaHSO_3$ and $Na_2S_2O_3$ are common reductants used to destroy excess oxidants. Saturated $CuSO_4$ solution is very specific for removing pyridine. It forms a blue-purple complex and acts as its own indicator, since the $CuSO_4$ solution is deeply colored when pyridine is present. Final washes can be done with brine to remove any traces of water from the organic phase.

The lessons of solvent extraction from beginning organic chemistry classes should not be forgotten by the aspiring synthetic organic

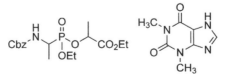

Figure 10.1 Examples of quite polar compounds that are extracted from aqueous reaction mixtures using $CHCl_3$/isopropanol mixtures.

chemist. While modern separation tools are amazing and powerful, one should not be seduced by the sophistication of the new. All too often, novices use a high-cost, labor-intensive technique like chromatography to separate two components in a reaction mixture (e.g., a phenol and a neutral) that could be easily separated by solvent extraction. A nice example of how compounds can be manipulated in solvent extraction via their acidity is seen in the hydrolysis of a 2,6-dimethylphenyl ester (eq. 30). Alkali initially produces the conjugate bases of both the acid and the phenol in an aqueous phase. If acidified with strong acid, both bases would be protonated and would be extracted into the organic phase. However, if CO_2 (in the form of ground dry ice) is added to the alkaline solution, it effectively converts the hydroxide to carbonate, which is only basic enough to ionize the acid, not the phenol. The phenol is removed by extraction into an organic phase, and the aqueous phase is acidified with HCl to recover the acid by extraction into a different organic phase.

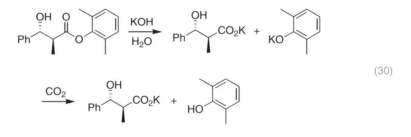

(30)

Problems can and often do arise in solvent extraction. For example, when using extraction with base to move an acid to the aqueous phase, its salt may be quite insoluble even though it is ionic. This event is difficult to anticipate; likely the only way to avoid this problem is to do a small-scale test extraction before the whole reaction mixture is extracted. Strategies to overcome the problem lie mainly in trying different cations (sodium → potassium, lithium, or ammonium). If this fails, the alternatives are to isolate the crystalline salt, or simply acidify the whole mixture to go back to ground zero. Keep in mind that salts of organic compounds are essentially soaps, and can facilitate the extraction of neutral organics into the aqueous phase. It may be possible to back-extract the aqueous phase to pull these neutrals back into an organic solvent before the aqueous extracts are combined for neutralization.

Likely the foremost difficulty in extractions is the formation of emulsions. Emulsions are most likely to occur with solvents of high viscosity and a low surface tension, and when there is a small density difference between the phases. These property data for a selection of solvents are collected in Table 7. Some are volatile solvents that would be used for extraction, and some are reaction solvents that may be present in the mixture. If it is the reaction solvent, not the extraction solvent, that seems to be creating the difficulty with the emulsion, an obvious approach is to change the problem solvent or remove it, by evaporation if possible, before extraction. Other tactics to deal with an emulsion include centrifugation, or filtration of the whole *gemische* through a pad of Celite® (trade name for diatomaceous earth). Celite is a filter aid composed primarily of silica, which can remove very fine particles that would otherwise clog a filter. Ultrasonication (in an Erlenmeyer flask) might also be worth a try. Increasing the difference in density between the phases or increasing their surface tension are other approaches. This can be done by dissolving a neutral electrolyte like Na_2SO_4 or $NaCl$ in the aqueous

Table 7 Selected Solvent Properties (at 25°C) that Can Affect Emulsion Formation

Solvent	Density $(g \cdot mL^{-1})$	Surface tension $(mN \cdot m^{-1\dagger})$	Viscosity $(mPa \cdot s^{\ddagger})$
Acetonitrile	0.786	28.66	0.369
Chloroform	1.492	26.67	0.537
Dichloromethane	1.325	27.20	0.413
Diethyl ether	0.706	16.65	0.224
Dimethylsulfoxide	1.100	42.92	1.987
Ethanol	0.789	21.97	1.074
Ethyl acetate	0.902	23.39	0.423
Hexane	0.672	17.89	0.300
Methanol	0.791	22.07	0.544
Pentane	0.626	15.49	0.224
Pyridine	0.978	36.56	0.879
Tetrahydrofuran	0.889	NA	0.456
Toluene	0.865	27.93	0.560
Water	1.000	71.99	0.890

$\dagger mN \cdot m^{-1}$ was formerly dyne $\cdot cm^{-1}$; $\ddagger mPa \cdot s$ was formerly cP; NA—not available.

phase or by adding a more polar, less dense solvent (ether, ethyl acetate, pentane) to the organic phase.

10.2 DRYING ORGANIC SOLUTIONS

After an aqueous extraction, the organic phase will include some dissolved water, so its drying is necessary before solvent evaporation. Four agents are commonly used for this purpose: $MgSO_4$, Na_2SO_4, K_2CO_3, and molecular sieves. A complete listing of desiccants is provided in Table 8. The choice can be based on the following considerations. It is easy to tell when $MgSO_4$ has fully dried an organic solution because dry $MgSO_4$ is fluffy but the hydrate is clumpy. A powerful drying agent such as $MgSO_4$ may be necessary for the somewhat polar and hydrophilic ether and ethyl acetate. While $MgSO_4$ is fast, it is also acidic, and acidic enough to remove ethylene ketals. For example, the apparently selective formation of the ketal of the saturated ketone in androstenedione (eq. 31) is in fact ketalization of both carbonyl groups and a rapid hydrolysis of the α,β-unsaturated ketal promoted by the $MgSO_4$ drying agent. If a compound is acid-sensitive, K_2CO_3 can be substituted for $MgSO_4$. K_2CO_3 is granular and free-flowing when dry but clumpy when wet, and of course is slightly basic. If a compound bears base-sensitive groups, like acetate esters, and a protic solvent is present, removal of the acetate might be expected. Na_2SO_4 likewise is granular and free-flowing when dry but clumpy when wet. It is also quite neutral, but it tends to be slower than either $MgSO_4$ or K_2CO_3, such that overnight drying of organic solutions is often required unless they are quite nonpolar and nonhydrophilic, like dichloromethane. Molecular sieves can also be used, but they offer none of the self-indicating features of any of the chemical drying agents. Following drying, filtration into a flask suitable for vacuum (round-bottom or pear-shaped) prepares for evaporation of the organic solvent.

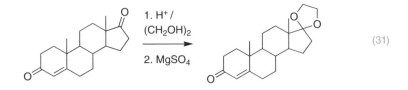

(31)

Table 8 Chemical Desiccants

Agent	Used with	Not used with	wt % water[a]	Mechanism
Al_2O_3	hydrocarbons		20	Adsorption
$CaCl_2$	halides, hydrocarbons, esters, ethers	alcohols, aldehydes, amides, amines, ketones	30	Hydration
CaO	alcohols, amines, $NH_3(g)$	acidic compounds, esters	30	Chemisorption
MgO	alcohols, aldehydes, amines, hydrocarbons	acidic compounds	50	Hydration
$MgSO_4$ (anh)	most compounds, incl. acids, aldehydes, esters, ketones, nitriles	none	ca. 50	Hydration
P_2O_5	anhydrides, ethers, halides, hydrocarbons, nitriles	acids, alcohols, amines, ketones	50	Chemisorption
K_2CO_3 (anh)	alcohols, amines, esters, ketones, nitriles	acids, phenols	20	Hydration
KOH (pellet)	amines	acids, phenols, esters	NA	Hydration
Silica gel	most compounds		20	Adsorption
Na_2SO_4 (anh, granular)	acids, aldehydes, halides, ketones		120	Hydration

[a] (g/100 g desiccant).

10.3.1   Reactions Producing Triphenylphosphine Oxide

Such reactions include Wittig and Mitsunobu Reactions. The following method will work only if the product is relatively nonpolar. The reaction mixture is concentrated, the residue is suspended in pentane or hexanes and a minimum amount of ether, and this solution is loaded onto a short "plug" (5 cm in height) of silica gel in a chromatography column. The product is eluted with ether, leaving most of the phosphine oxide on the silica gel. Sometimes it is necessary to repeat this procedure two to three times.

10.3.2   Reactions Involving Boron Compounds

Many boron compounds and boron-containing residues (e.g., from Suzuki coupling, $NaBH_4$ or B_2H_6 reduction, or hydroboration) can be removed from a reaction mixture by evaporating it repeatedly from MeOH. This process forms $(MeO)_3B$, which is volatile and therefore also removed.

10.3.3   Reactions Involving Copper Salts

Such reactions include those using organo cuprates. Quench the reaction with a saturated solution of aqueous NH_4Cl and allow the mixture to stir for a few hours while open to the atmosphere. This will oxidize all cuprous salts to the cupric oxidation state, which is much more reactive to ligand exchange. This will permit complexation of the cupric ion by ammonia, as indicated by the dark blue color of the solution.

10.3.4   Reactions Involving Aluminum Reagents

Such reactions classically involve $LiAlH_4$, but this method may be more broadly applied. The aluminum hydroxide formed upon aqueous quenching can create impressive emulsions. This material can be dissolved in aqueous acid, but not all reaction products can tolerate this, and the product losses into the aqueous phase can be substantial. A method that avoids adding large amounts of acid to work up $LiAlH_4$ reductions uses the following method, called the "Fieser" or N,N,3N workup. The reaction mixture is carefully quenched (H_2 evolution!) with 1 mL of H_2O for each gram of $LiAlH_4$ used. After a few minutes of stirring, 1 mL of 15% NaOH for each gram of $LiAlH_4$ is added, and after another few minutes of stirring, 3 mL of H_2O for each gram of $LiAlH_4$ is added. This sequence converts the aluminate salts into alumina, which should be an easily filtered, granular material. No further aqueous workup is necessary. A drying agent can be added if desired, and filtration and evaporation provide the product.

10.3.5   Reactions Involving Tin Reagents

Such reactions include the free radical reduction of halides. Sodium fluoride solution is useful for the workup of reactions of Bu_3SnH, which works by the production of insoluble fluoride salts that are removed by filtration (Caddick, Aboutayab, and West, 1993). Stille couplings also produce triallcyltin compounds and may be purified by similar methods.

CHAPTER

11

EVAPORATION

The rotary evaporator (Fig. 11.1) is an essential piece of equipment in most organic laboratories. Commonly vacuum is provided by a water aspirator (or by other mild vacuum sources), and the flask is rotated by an electric motor. This provides a thin film of solution for evaporation, hopefully preventing the solution from bumping. The flask rotates in a temperature-controlled water bath, which provides the heat of vaporization. Little heat may be needed to evaporate highly volatile solvents like ether, whereas heat is definitely required for solvents like toluene. It is essential that a glassware bulb (Fig. 11.2) be inserted between the flask and the evaporator to make provision for bumping of the solution being evaporated. If a bump does occur, the solution is trapped in the bulb, enabling it to be transferred back into the flask. Without the bulb, the solution would be drawn up into the evaporator itself, where it could mix with condensed solvents, be contaminated by other compounds that have been earlier evaporated in the equipment, or be lost altogether. This bulb naturally must be rinsed with solvent after evaporation of each sample. The flasks used on a rotary evaporator may be of many sizes and types, but typically round-bottom or pear-shaped flasks are used. Flat-bottom flasks are not recommended because they are less robust to vacuum. The evaporator generally has a condenser or trap (circulating cold water or dry ice) that can condense evaporated solvents to prevent their being drawn into the waste water or vacuum source and/or to potentially permit solvent recovery and recycling.

Few beginning chemists have avoided the following embarrassing scenario: a flask was fitted onto the joint on the rotary evaporator, the vacuum and motor were turned on, and the chemist walked away, having forgotten to close the vent to the vacuum system and wait for the vacuum to take hold to secure the flask. The friction fit of the flask on the joint of the glassware trap may have held it for a time,

The Synthetic Organic Chemist's Companion, by Michael C. Pirrung
Copyright © 2007 John Wiley & Sons, Inc.

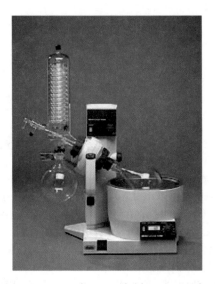

Figure 11.1 The rotary evaporator is essential in any synthetic organic chemistry lab and is the scientific instrument most fascinating to lay reporters. It is invariably the backdrop for any television or print news story on chemistry. Photo provided by Brinkmann.

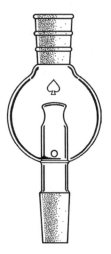

Figure 11.2 The rotovap bulb prevents bumps from solvent in the evaporating flask from being carried into the rotary evaporator.

but eventually it slipped off into the water bath. As frustrating as this situation is, all is not lost. It is absolutely possible to perform solvent extraction on the water in the rotovap bath to recover at least some of the product. The yield obtained in today's reaction is not something one could quote for publication, however. Considering the effort that goes into carrying synthetic intermediates through several chemical steps, the chemist simply cannot afford to allow this loss of material without some effort at recovery. This experience also serves as a useful lesson that most never forget.

The size of the flask used in evaporation should ideally be at least twice the volume of the solvent. Often, this means a fairly large flask even if the amount of product is 100 mg or less. Such large flasks may not even fit onto the analytical balance used to obtain an accurate mass of the reaction product. Typically, an initial evaporation of the large volume is followed by transfer of the residue into a smaller, tared flask. This transfer is conveniently done using a long Pasteur pipette and a rubber bulb. A few mL of a nicely volatile solvent like ether or dichloromethane are used to dissolve the residue by squirting the solvent down the walls of the flask while taking care to avoid the ground glass joint. The resulting solution is then drawn into the pipette and transferred into the smaller flask. Repetition of this process once or twice more should ensure quantitative transfer.

One is sometimes faced with the task of transferring a viscous material into a small container, such as a vial. Dissolving it in a solvent is necessary, but evaporation in the vial is a challenge. An adapter can be fitted with a rubber stopper that snugly fits into the mouth of the vial, and rotary evaporation can be applied.

Centrifugal evaporators (Fig. 11.3) are relatively new to the chemistry laboratory, having made their way over from the biology laboratory. They were initially designed for the vacuum evaporation of ethanolic DNA solutions held in plastic tubes. Many different rotors are now available to accept many types of laboratory containers in large numbers. Centrifugal force keeps the liquid within each container and prevents bumping losses. This instrument is essential to high-throughput, parallel approaches to organic synthesis.

Rotary evaporation is rarely sufficient to remove all traces of solvent, especially if a higher boiling solvent like toluene or a solvent with a high heat of vaporization like *tert*-butanol is present. It is always advisable to place the evaporation flask under high vacuum

Figure 11.3 A centrifugal evaporator. Tomy Seiko and Subsidiaries.

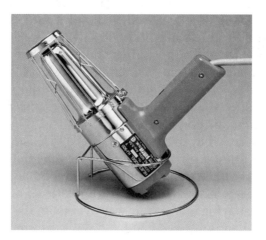

Figure 11.4 A heat gun.

(<1 torr). With solvents that are more difficult to evaporate, this high vacuum may still prove insufficient to remove the last of it, even when kept under vacuum overnight. A quick application of heat from a heat gun (Fig. 11.4) will frequently show the telltale signs of evaporation in the form of bubbling. At this stage the product should show little residual solvent by NMR.

It is good practice to save a TLC-sized sample of the crude reaction mixture (after evaporation but before any purification) and to record its NMR spectrum. These data can assist in understanding a purification by column chromatography, since TLC and NMR data will be obtained on each fraction. It sometimes appears by TLC that the major component of the reaction mixture has been isolated, but the NMR of the crude reaction mixture reveals that it is in fact a minor component and the major component is still to be found. The ratios of starting material to product (percent conversion) are available from crude NMR data, as are the ratios of product(s) to by-product(s) and the ratios of any isomers present. It is particularly important to determine isomer ratios before any fractionation of the reaction mixture, since it is possible that one of the isomers could be selectively lost during purification, thereby perturbing the actual ratio that was obtained in the reaction. Crude TLC and NMR data are also important to understanding all the products generated during a reaction, not just the major product, which is helpful during reaction optimization.

It is also important to obtain the mass of the crude reaction product after evaporation. While a yield cannot be calculated until the product has been purified, knowing how much of the mass that was added to the reaction mixture has been returned in the crude product ("the mass balance") can be very useful in troubleshooting reactions. If the mass balance is low, for example, the workup may have been poorly designed to recover the desired product. If the mass balance is far above the expected value, the solvent may have polymerized. The mass balance is simply determined by evaporating the crude reaction mixture in a tared flask. Since evaporation flasks are either round-bottom or pear-shaped, they must be prevented from tipping over or rolling across the bench. Cork rings or rubber filter adapters are available in sizes appropriate to all size of flasks and do an excellent job. A visit to any synthesis lab will reveal benches with many flasks containing products. It is crucial to label these flasks to avoid mix-ups. Adhesive labels could be used, but they change the tare of the flask. Better are tags with short loops of string that can be wrapped around the neck of the flask.

CHAPTER
12
VACUUM SYSTEMS

12.1 VACUUM SOURCES

Vacuums are obtained in the laboratory in at least three ways: "house" vacuum, water aspirators, and vacuum pumps. The strength of the house vacuum is defined locally. Water aspirators should be capable of providing a vacuum of about the vapor pressure of water. This is around 24 torr at room temperature; colder water will translate to a better vacuum. This pressure is quite adequate for many operations, such as rotary evaporation and some distillations. A trap bottle should be inserted between the aspirator and the apparatus, consisting of a thick-walled Erlenmeyer filter flask bearing a rubber stopper with a hose connection and a stopcock (for releasing the vacuum). Its role is to prevent water from entering the apparatus in case of a backup at the aspirator. Long (15 cm) tubing extending downward from the aspirator is necessary to obtain optimum vacuum. Some chemists object to aspirators because they waste water. For the green chemist, aspirator pumps that recirculate the water are available.

A common type of laboratory vacuum pump uses an electric motor and belt-driven rotary oil pump (e.g., Welch 1400 Duo-Seal; Fig. 12.1). These pumps are capable of vacuums of 0.1 torr (100μ) or better when operated with manifolds and other apparatus. They are used for vacuum lines, inert atmosphere boxes, distillations, and the like. Another type of vacuum pump has direct drive between the motor and the oil pump and is fully integrated. These are commonly used with instruments like mass spectrometers because of their low maintenance.

The Synthetic Organic Chemist's Companion, by Michael C. Pirrung
Copyright © 2007 John Wiley & Sons, Inc.

Figure 12.1 A conventional laboratory oil vacuum pump. Photograph provided by Welch.

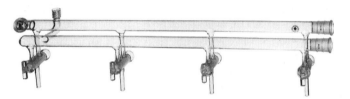

Figure 12.2 A glass double manifold for simultaneous manipulation of the atmosphere in a vessel, including vacuum and delivery of an inert gas.

12.2 VACUUM MANIFOLDS

Working at reduced pressure is significantly easier when a glass vacuum manifold is used (Fig. 12.2). It may also be used with (or even integrated with) the inert gas manifold discussed earlier. Components of the system often include a pump connection, a cold trap (to condense organic vapors, preventing them from being drawn into the pump), a central chamber with several branching stopcocks (best are hollow vacuum stopcocks whose interior can be evacuated to keep the stopcock seated; Fig. 12.3), and a pressure gauge. It is essential in any vacuum system that one stopcock always remains available to the ambient atmosphere to permit release of the vacuum. A vacuum pump must be vented to the atmosphere when turning it

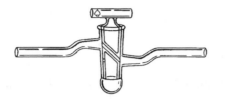

Figure 12.3 A hollow plug vacuum stopcock is the stopcock most resistant to leakage.

off; otherwise, oil will be drawn into the manifold system. If liquid N_2 is used to cool the trap, it should not be applied until the system has mostly been evacuated. Otherwise, liquid oxygen can be condensed in the trap, which can cause an explosion. The ideal trap design has stopcocks both before and after the trap as well as a stopcock to vent it to the atmosphere. This arrangement allows the trap to be brought to atmospheric pressure for emptying without venting the rest of the vacuum system.

When a vacuum system is not performing well, the following items should be checked. These are also good maintenance practices. Although there are no general rules for how often these measures need to be taken, monthly checks of these items are advised.

1. Change the pump oil. Run the pump for 15 minutes to warm the oil before draining it. Add new oil slowly to give it a chance to flow until it appears at the correct level, typically in a sight glass. If the removed oil seems particularly dirty, it may be worth using a "flushing oil" first. These materials are actually hydrocarbons that are less viscous than oil. Run the pump for 15 minutes against a closed system, drain the flushing oil, and replace it with the normal (high-vacuum) oil. Repeat the foregoing with high-vacuum oil. Simple replacement of the high-vacuum oil is the normal maintenance procedure.

2. Check that all ground glass joints are fully seated and well-greased with a good high-vacuum grease like Apiezon M. The same goes for stopcocks, except the grease should be Apiezon N.

3. Minimize the lengths of hoses and the numbers of joints and stopcocks.

4. Fill the cold trap with a coolant such as dry ice/isopropanol or liquid N_2.

12.3   VACUUM GAUGES

It should be emphasized that while it is crucial that the pressure be reported for some operations like vacuum distillation, there are significant errors in most simple measurements of vacuum.

At least two types of McLeod gauges (swivel and tipping) are available. They have a rest position (when they are exposed to the vacuum) and a reading position. With the swivel gauge (Fig. 12.4), the mercury reservoir is down at rest. The gauge has both linear and

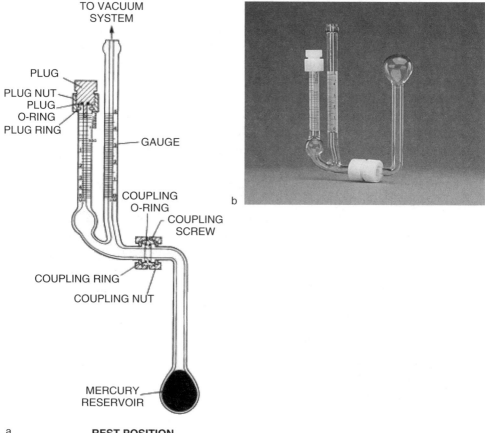

a   **REST POSITION**

Figure 12.4 The swivel McLeod pressure gauge, in schematic (*a*) and photographic (*b*) forms. © Sigma-Aldrich Co.

nonlinear scales, with the latter for higher accuracy at pressures less than 1 torr. To read the pressure, the reservoir is swiveled forward and up past the horizontal, until the mercury level in the closed capillary reaches the lowest line as shown (Fig. 12.5). Pressure is read from the level of the mercury in the open capillary. If the pressure is less than 1 torr, the nonlinear scale (graduated in microns [μ, 10^{-3} torr]) is used as follows: continue to swivel the reservoir up until it reaches vertical, then carefully adjust the mercury level in the open capillary until it is at the top line. The level of the mercury column in the closed capillary provides the pressure.

Tipping McLeod gauges (Fig. 12.6) measure pressures from around 5 torr to 5 microns. In the rest position the glass portion is turned 90° to the right from the position shown, with mercury filling the pear-shaped reservoir. To read the pressure, the glass portion is rotated to the left (to the position shown) and mercury enters the closed arm. The pressure is given on the scale on the closed arm.

Mercury manometers (Fig. 12.7) are simpler and cheaper than McLeod gauges, though they cannot measure pressures lower than

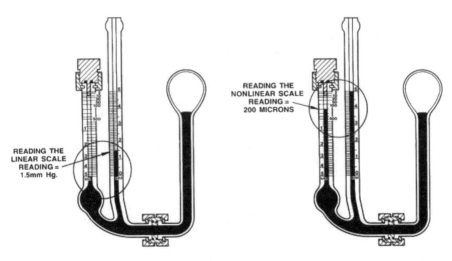

READING THE LINEAR SCALE   READING THE NONLINEAR SCALE

Figure 12.5 Reading the swivel McLeod gauge on the linear and nonlinear scales. © Sigma-Aldrich Co.

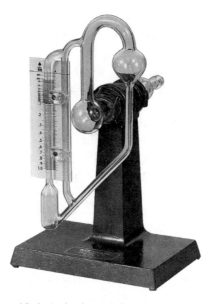

Figure 12.6 A tipping McLeod pressure gauge.

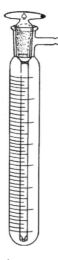

Figure 12.7 A mercury manometer.

a few torr. They are often used in conjunction with aspirators. Like many manometers, pressure readings are based on the difference in the heights (in mm, translated to torr) of the inner and outer mercury columns.

SAFETY NOTE

Manometers are the most likely source of larger mercury spills in the laboratory, but broken thermometers are the more frequent source. Mercury metal is a toxic, volatile liquid. Mercury vapor is readily absorbed by inhalation and can also pass through intact skin. Mercury is highly hazardous when inhaled or when it remains on the skin for more than a short period of time, commonly resulting in penetration to the central nervous system and mercury poisoning. Elemental mercury is not well absorbed by the gastrointestinal tract. Consequently it is only mildly toxic when ingested. Mercury spills generate an enormous number of tiny droplets that are easily dispersed, and special vacuum equipment must be used to clean up these spills. Do not touch or attempt to wipe up spilled mercury. The safety office or other environmental services can be called to safely and properly clean up the spill. Make sure someone stays near (but not in) the spill zone to keep people away.

CHAPTER
13
PURIFICATION OF PRODUCTS

If the methods described so far have not produced a single pure compound, as evidenced by the number of spots on TLC, the number of peaks on GC, or an NMR spectrum, purification must be undertaken.

13.1 DISTILLATION

Kugelrohr distillation serves only to separate nonvolatile materials from volatiles and is always performed under reduced pressure, which keeps the glassware assembled while in use. This technique can be performed on a relatively small scale (hundreds of milligrams) without serious material losses. It is also sometimes called bulb-to-bulb distillation (Fig. 13.1). One "bulb" (round-bottom flask) is held in an electrically heated oven and is attached through a hole in the oven wall to an outside bulb (Fig. 13.2). This latter bulb is attached by a hose adapter and rubber tubing to a pneumatic drive motor that reciprocally rotates the glassware, like a windshield wiper, to prevent "bumping" (sudden violent eruptions of vapor that carry liquid over into the receiver). The motor drives a hollow shaft through which the vacuum is pulled. The external bulb is cooled (with an ice/H_2O or dry ice/acetone bath) to condense the vapor. Once the compound has distilled, it may be removed from the bulb by pouring it out (if it is free-flowing) or washing it out with ether.

The short-path distillation method described for the purification of reagents may also be applicable to product purification, provided that the scale is toward the larger, 25 mmol end of the spectrum that is the main focus of this book. The number of compounds for which distillation is applicable is somewhat limited, but when it is, distillation can be an excellent purification method. Compounds purified by chromatography may remain tenaciously colored despite

The Synthetic Organic Chemist's Companion, by Michael C. Pirrung
Copyright © 2007 John Wiley & Sons, Inc.

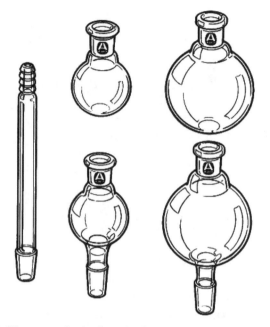

Figure 13.1 Glassware for bulb-to-bulb distillation. © Sigma-Aldrich Co.

appearing quite pure by other criteria, whereas distilled compounds are most often water-white (provided they have no chromophore).

13.2   SILICA GEL CHROMATOGRAPHY

Silica gel chromatography is by far the most widely used method of product purification because it is so general and because it is a natural outgrowth of the analysis of reaction mixtures by TLC. For the most part, solvent mixtures that are effective for TLC can also be used for column chromatography, with some adjustment of proportions, mostly to reduce the R_f. One exception is that acetone should not be used in the elution of a preparative column because it easily undergoes aldol condensation to give 4-methylpenten-2-one.

For many years, column chromatography was performed with gravity elution that might require hours. Automated fraction collection was essential to permit columns to run while the chemist performed other tasks. This all changed with the development of

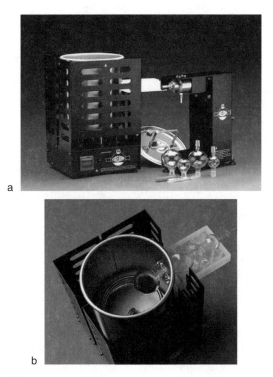

Figure 13.2 A Kugelrohr distillation oven. (*a*) Side view; (*b*) top view. © Sigma-Aldrich Co.

flash chromatography, which has many virtues, chief among them being much faster separations (Still, Kahn, and Mitra, 1978). Gas pressure is used to push eluent through silica gel with a small pore size; separation is generally complete in minutes, meaning that fractions can be collected manually in test tubes in a rack. It is thus a "poor man's" HPLC.

13.3 FLASH COLUMN CHROMATOGRAPHY

Since its original development, many proper and improper modifications have been made in or incorporated into the practice of flash chromatography by novice synthetic chemists. It is therefore worthwhile to return to the original publication to review the first principles of this widely used method. First, it uses 40 to 63 μm silica gel,

Table 9 **Recommended Parameters for Flash Column Chromatography at Different Scales**

sample loading (mg)	column diameter (mm)	fraction size (mL)
100	10	5
400	20	10
800	30	20
1600	40	30

[Still, Kahn, & Mitra, 1978].

which had been used for thin-layer chromatography. Still (1978) recommends selecting an eluting solvent that gives an R_f on TLC of the target compound of 0.35. It is important to note the difference between this recommendation and that for analytical TLC. He also recommends column diameters for particular separation scales (Table 9) and that the pressure be adjusted so that the solvent head drops at the rate of $5 \, cm \cdot min^{-1}$. Still claimed the ability to separate compounds whose R_fs differ by 0.15, and in some fortunate cases, 0.1. Past studies on gravity flow columns have reported separations of compounds whose R_fs differ by 0.01. So one limitation of the flash chromatography method is reduced resolving power. Having experienced the rapid separation of a flash column, however, chemists are unlikely to return to gravity flow columns to achieve separation of close-running compounds. Preparative chromatography instruments available today are much more likely to be used for this task.

Flash chromatography uses a column with a Teflon stopcock and a solvent reservoir (Fig. 13.3). The reservoir ideally stores sufficient eluent to complete the chromatographic separation so that the pressure need not be released to replenish the solvent. The column is packed dry instead of being slurry packed like a gravity column. A valve consisting of a Teflon screw valve, a 24/40 joint, and a gas inlet is placed on the top of the column and attached to a source of compressed air. Some chemists firmly attach the valve to the column and adjust the flow rate with the air pressure and the screw valve. Others close the screw valve completely, turn on the compressed air, and use the force with which the valve is pressed onto the top of the column (by hand) to control the pressure and therefore the flow rate. The reservoir is loaded with solvent, and it is pushed through the column. As expected, the mixing of silica gel with the solvent

Figure 13.3 A flash chromatography column, reservoir, and air control valve.

produces heat that cracks the silica gel bed, but this is not a problem because continued pushing of solvent through the bed drives out air bubbles and packs the silica gel uniformly. After the column is equilibrated and the solvent is pushed through until it is just at the top of the silica packing, the column is topped by sand and the sample is added. It can be added in solution, or it can be dissolved in a volatile solvent that is mixed with silica gel and evaporated to give a powder.

Flash chromatography can be conducted on a very small scale, and is more useful than other small-scale chromatographic techniques like preparative TLC. A Pasteur pipette can be used as a flash column, for example. Glass wool is pressed into its narrow neck and covered with a layer of sand. Silica gel is added to about half the height of the main barrel, the pipette is clamped to a rack, and eluting solvent is added above. Tygon tubing connected to a compressed air source is slipped over the top of the pipette to provide pressure. Fractions are collected in a rack of small vials.

13.4 GRADIENTS

Despite some similarities to HPLC, column chromatography differs from it in that continuous solvent gradients are rarely used, as they are difficult to generate. In some cases step gradient elution may be used, for example, involving several column volumes of 10% solvent A in solvent B, next 30% A in B, then 70% A in B. In flash chromatography, this technique requires release of the pressure to change solvents in the reservoir, whose inconvenience may inhibit the use of gradients.

13.5 SPECIAL ABSORBENTS

13.5.1 Triethylamine-Treated Silica Gel

Compounds that are sensitive to the acidic character of conventional silica gel can sometimes be separated if the silica gel has been pretreated with a deactivating agent. An eluent that includes 1% by volume of triethylamine may be used in the initial wetting and equilibration stage preparing for flash chromatography. Column elution may be conducted with or without triethylamine in the eluent. Laboratory mates will appreciate elution without triethylamine because of its base odor. Examples of sensitive compounds that can be purified using this strategy include nucleoside phosphoramidites (Fig. 13.4).

13.5.2 Oxalic Acid-Coated Silica Gel

This material was developed for the purification of particularly sensitive quinones (Fig. 13.5), (Yamamoto, Nishimura, and Kiriyama, 1976) and has been used in flash chromatography (Pirrung et al., 2002). It is prepared by suspension of silica gel in 0.1 N oxalic acid

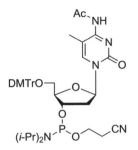

Figure 13.4 A nucleoside phosphoramidite can be purified by chromatography on triethylamine-treated silica gel.

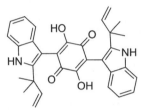

demethylasterriquinone B4

Figure 13.5 A quinone whose purification by chromatography requires oxalic acid-coated silica gel to avoid decomposition.

overnight, filtration, washing with H_2O, and drying in an oven at 100°C overnight.

13.5.3   Silver Nitrate-Impregnated Silica Gel

This sorbent was developed long ago to address the issue that hydrocarbons generally have little affinity for polar silica gel (Williams and Mander, 2001). Therefore hydrocarbons migrate with a high R_f on conventional silica gel sorbents regardless of the eluting solvent, and no separation of hydrocarbons can be achieved. An approach was sought to increase the interaction of hydrocarbons with the stationary phase and form the basis of separations. Many hydrocarbons, including those that are naturally occurring, include $C=C$ bonds, and it was known that such bonds have a high affinity for metal ions, particularly silver. The idea then was that silver ion would be added to the stationary phase to increase the retention of alkenes.

α-pinene   β-pinene

Figure 13.6 The terpenes α-pinene and β-pinene can be easily separated only on silver nitrate-impregnated silica gel.

Not only did this achieve their separation from alkanes, differently substituted alkenes could be separated from one another based on the strength of their retention by the stationary phase, as alkene affinity toward silver ion is dependent on substitution. It is thus possible to separate the turpentine constituents α-pinene and β-pinene (Fig. 13.6). Protocols for preparation of silica columns including silver ion were therefore developed. These primarily involve dissolving silver nitrate in methanol or acetonitrile, adding it to the silica gel, and evaporation and oven drying.

Access to silver nitrate-impregnated TLC plates is essential to making the decision to use silver nitrate chromatography. They must also be used in selection of solvents and analysis of column fractions. Getting and using these plates is somewhat problematic because the photoreactivity of silver ion makes advance preparation and storage of silver-impregnated silica gel difficult. To avoid the need for preparation of home-made TLC plates including silver, methods have been developed to modify commercial TLC plates with silver (Ratnayake, 2004). Soaking them in 12.5% (wt/vol) aqueous $AgNO_3$, or allowing this aqueous solution to be drawn up the plate as if one was performing TLC, adds the required metal ion. Spraying of the $AgNO_3$ solution onto the plate is less desirable because it is not uniform. Water deactivates the silica gel, so the silver TLC plates must then be reactivated by heating at 80–110°C for 1 to 2 hours. These operations and storage of the plates are performed with the exclusion of light to the extent possible.

13.5.4 Other Sorbents

Other stationary phases that can be used for column chromatography include florisil, which is magnesium silicate, and alumina. Broadly speaking, their absorptive properties are greater than silica gel.

13.6 PREPARATIVE GAS CHROMATOGRAPHY

Packed column GCs may still be found in some laboratories. With relatively large diameter columns, these instruments can accept large amounts (up to 50 µL) per injection. With a thermal conductivity detector (TCD), the analysis is nondestructive, allowing preparative separations. This basically involves monitoring the recorder to identify when a desired compound is eluting. A 3 mm glass tube or a specially made collector is attached at the exit port and the vapor flowing off the GC column is condensed. Some compounds condense better when the collector is cooled (ice or dry ice), while others form aerosols with cooling and are better collected in a warmer container (wrapped with aluminum foil). After collection, one end of the glass tube may be sealed and the liquid centrifuged to the sealed end. This sample is now ready for mass spectrometry or combustion analysis.

CHAPTER

14

METHODS FOR STRUCTURE ELUCIDATION

Many useful and detailed texts concerning the elucidation of organic structures using spectroscopic data are available. This chapter aims to address only practical issues in collecting the data.

14.1 NUCLEAR MAGNETIC RESONANCE SPECTROSCOPY (NMR)

Proton NMR requires a nonprotonic solvent, generally deuterated, to avoid obscuring signals of the desired compound by solvent signals. An internal chemical shift standard, such as TMS (tetramethylsilane), may optionally be included in the solvent. The cheapest organic NMR solvent is $CDCl_3$, because it can be prepared by base-catalyzed exchange in D_2O (eq. 32). It is never 100.0 atom % deuterium, however, so a small singlet from residual protons in the solvent (i.e., $CHCl_3$) is often seen at around 7 ppm. Some workers use this signal as the internal standard. It should also be kept in mind that an intrinsic property of chloroform is that it undergoes slow decomposition to HCl, which will also occur (to DCl) in $CDCl_3$. If sensitivity to acid of the compound under analysis is a concern, another NMR solvent should be chosen. Another commonly used solvent is d_6-acetone, which like chloroform is volatile and permits the easy recovery of the compound from the NMR sample. Other solvents that are often used for polar or ionic species include d_6-DMSO and D_2O, which are not easily evaporated. Essentially any other solvent is available in deuterated form, the main issue being the expense.

$$CHCl_3 \quad + \quad D_2O \quad \overset{\ominus OD}{\rightleftharpoons} \quad CDCl_3 \quad + \quad HOD \qquad (32)$$

When using a Fourier transform NMR instrument, 1 mg of compound is sufficient to obtain a quite reasonable proton NMR

The Synthetic Organic Chemist's Companion, by Michael C. Pirrung
Copyright © 2007 John Wiley & Sons, Inc.

spectrum. Carbon NMR requires a much larger quantity of compound than proton NMR, because of the low natural abundance (1.1%) of ^{13}C. To prepare the sample, dissolve the compound in 1 mL of the deuterated solvent, which should be enough to fill the NMR tube to about 5 cm depth. If a smaller volume is being used, it is essential to a create a cylinder of constant magnetic susceptibility within the RF coils of the probe. This is accomplished with a susceptibility plug set (Fig. 14.1) or a Teflon vortex plug. The vortex plug comes with a matching threaded metal rod. The plug is gently pushed into the tube with the rod until the bubble of air above the solution disappears. The tube is then capped. After the spectrum is collected, the plug is removed by threading the rod into the plug and pulling, very gently.

It is important to remove solids or particulate matter from the sample, which will affect the magnetic field homogeneity and reduce the resolution of the spectrum. The sample can be filtered through a glass wool plug in a Pasteur pipette, with the filtrate flowing directly into the tube. Dissolved metallic and paramagnetic impurities (e.g., Fe^{3+}, Mn^{2+}) are another factor that can cause reduced resolution. It is difficult to remove such species at sample preparation time. All NMR tubes must be sealed with plastic caps. Color coding based on the cap will help keep track of samples when taking several spectra at one sitting. Some workers use paper tags looped onto their tubes with string for this purpose.

An NMR tube adequate for routine spectra is the Wilmad 6 PP (5 mm) tube. For important, high-quality spectra, the Wilmad 7 PP tube can be used. Chipped, cracked, or broken tubes should not be used, but they might be resurrected by sanding off the broken part using a glass sanding wheel, if the glass shop has one. Tubes may be cleaned by rinsing with ether and acetone. It is important not to use

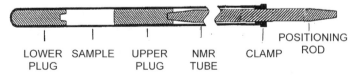

LOWER PLUG   SAMPLE   UPPER PLUG   NMR TUBE   CLAMP   POSITIONING ROD

Figure 14.1 A Doty susceptibility matching plug set. Provided by Doty Scientific.

any kind of pipe cleaner or brush to clean the inside of an NMR tube. Scratches will lead to "blips" of sample outside the cylinder defined by the rest of the tube, causing inhomogeneity. KOH baths are not good for NMR tubes, and they should never be placed in any type of chromic acid, as even tiny deposits of paramagnetic metals will severely affect resolution. Aqueous solutions and salts are removed from NMR tubes with much difficulty because of their long, narrow internal dimensions and the surface tension and viscosity of aqueous solvents. Special apparatus are available that enable solvents to be shot to the bottom of an inverted tube and then drain (Fig. 14.2). Some workers dry NMR tubes in an oven, but it is best

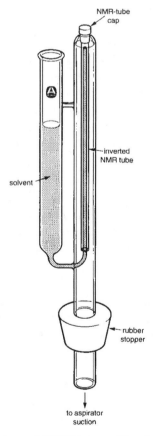

Figure 14.2 An apparatus to wash NMR tubes with solvents. © Sigma-Aldrich Co.

not to exceed 60°C, meaning that the same oven used for drying glassware cannot typically be used. This precaution is necessary because the thin glass may be distorted by heat. It can introduce curvature relative to the tube's long axis, causing the sample to wobble when spinning, break in the probe, or potentially damage the probe's radio frequency coil by contact. Tube distortion is also the major source of spinning side bands and increased shimming times.

When a compound has hydrogens bound to O, S, or N, their chemical shift and line shape can be quite variable, depending on parameters such as solvent, temperature, concentration, and impurities. It can be very useful to know which signals in the ^{1}H NMR spectrum correspond to those protons, and this can be determined by showing which signals are "exchangeable"—that is, which can move from molecule to molecule according to the equilibrium in (eq. 33). In this example, hydrogens are switched between the "blue" and the "red" molecules, which leads to a phenomenon called exchange broadening. The theoretical treatment of exchange is left to sophisticated NMR spectroscopy texts, but it has a practical impact on the appearance of the spectrum. An obvious way to examine exchange would be to add a deuterated, exchangeable molecule such as D_2O to the sample and observe the disappearance of the signal for the X–H group as it is converted to the X–D group (eq. 34). Typically the spectrum is acquired normally, the D_2O is added, and the spectrum is acquired again for comparison. Because D_2O is immiscible with most organic NMR solvents, vigorous shaking is required for mixing, and of course the heterogeneous mixture may affect magnetic field homogeneity and therefore spectral resolution. Another approach is to add a drop of formic acid (miscible with organic solvents) to the sample following acquisition of the first spectrum. Its protons are far downfield and therefore will not interfere with most signals in the spectrum. Its effect is to accelerate exchange (eq. 35), which should change the line shape of the exchangeable protons, and often shifts them downfield.

$$CH_3OH \quad + \quad CH_3OH \quad \rightleftharpoons \quad CH_3OH \quad + \quad CH_3OH \qquad (33)$$

$$CH_3OH \quad + \quad D_2O \quad \rightleftharpoons \quad CH_3OD \quad + \quad HOD \qquad (34)$$

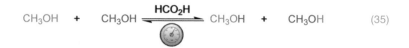

$$CH_3OH \quad + \quad CH_3OH \; \underset{}{\overset{HCO_2H}{\rightleftharpoons}} \; CH_3OH \quad + \quad CH_3OH \qquad (35)$$

It should also be emphasized that many NMR facilities have their own policies and recommendations for sample preparation and spectrum acquisition that may preempt the more general advice given here.

Comparing the spectrum one has taken today with a tabulated spectrum of the same compound from the literature or an actual spectrum from the several available spectral databases has its pitfalls. Chemical shifts can be quite solvent dependent, so if one is not using the same solvent in which the earlier spectrum was recorded, correspondence of peaks may be imperfect. Chemical shifts are also concentration dependent in some cases. Synthetic chemists almost never concern themselves with concentration in preparing an NMR sample, so this factor cannot be taken into account. NMR spectrum prediction software has recently become readily available on personal computers. While this tool can certainly be helpful, the algorithms used in these programs may not be perfect, and there is no substitute for an actual NMR spectrum for visual comparison.

The best source for a comparison spectrum is one that has been taken on the same instrument, which is available if NMR has been used to test reagent purity. To facilitate the direct comparison of starting material spectra with those of reaction products, the chemist should develop the habit of using the same plotting procedure for each spectrum. That is, always plot the proton spectrum between 0 and 8 ppm (or 10 ppm if aldehydes are frequently used) or the carbon spectrum between 0 and 200 ppm on a single sheet. With high-field NMR spectrometers that spread the proton chemical shift range over thousands of Hertz, these plots do lose some detail, but this information can be captured in subsequent "blowup" plots (small portions of the spectrum printed at a larger size) of specific regions of interest.

When examining NMR spectra of crude reaction mixtures, extraneous peaks may frequently appear owing to solvents remaining from the reaction or the purification, or from contaminants found within them (e.g., dioctyl phthalate or 2,6-di-*tert*-butyl-4-methylphenol). To facilitate identification of such peaks in high resolution NMR spectra, a comprehensive compilation was made by Gottlieb et al. (1999). These data are presented in Appendix 1.

14.2   INFRARED SPECTROSCOPY (IR)

There is a fair amount of variation in acquiring infrared spectra based on the particular spectrometer used. The reader is referred to the many textbooks, user manuals, and local instrumentation facility guides for further details. Infrared spectroscopy has also fallen into disuse in some laboratories, perhaps because of the amazing advances in and immense power of NMR. However, there are still questions about organic compounds that are best asked and answered with IR. Many research advisors have been frustrated by a student's conclusion that a compound has no carbonyl group based on the absence of protons in the 2 to 3 ppm region. An IR spectrum is by far the best way to make inferences concerning functional groups present in the molecule.

For liquids, spectra are typically obtained using thin films formed between two NaCl windows (also called salt plates). Free-flowing liquids can be dotted neat, using a Pasteur pipette, onto one salt plate as a single drop of as little as 1 mg. Holding the other plate by its edge, the first plate is covered with it and they are pressed together, spreading the sample. Provided the two plates have smooth surfaces, a film should be readily seen between them. This assembled sandwich is inserted into a holder for acquisition of the spectrum. If the sample is a viscous liquid or a solid, it is sometimes possible to form a thin film by transferring onto the plate a solution of the compound in a minimum volume of a volatile solvent (diethyl ether, dichloromethane). Evaporation of the solvent under a stream of nitrogen generates the thin film. After acquisition of the spectrum, the sample may be recovered by rinsing it off the plate with diethyl ether, and the plate is cleaned with acetone. Windows will inevitably become cloudy and their surfaces will roughen over time. They can be polished by dropping 50% aqueous ethanol onto a paper towel and moving the window in a circle on the towel, first on one side and then on the other. This wet solution should be removed with an acetone rinse and the plates dried under a stream of nitrogen. Storage in a desiccator is recommended.

While more laborious, infrared spectra can also be obtained in solution, usually in chloroform. $CHCl_3$ has a simple IR spectrum that does not interfere with most signals in the spectrum of the sample, and a reference cell can be used to eliminate those. A solution spectrum of a liquid sample may be obtained to examine bands whose appearance is concentration dependent, for example OH stretches. A solution spectrum of a solid sample may be obtained because the thin film technique described above failed. IR solution cells have two salt windows held a fixed distance apart, with the space between accessed through filling ports (usually Luer connectors) that can be plugged. Approximately 3 mg of the sample is placed in a vial and dissolved in 4 mL of solvent. The solution is taken up in a 1 mL syringe and injected into one of the two open ports of a solution cell. One plug is inserted, the cell is tilted to allow air to escape, and the other plug is inserted. If available, a reference cell is filled with pure solvent in the same way. Sample recovery can be accomplished by flushing the solution out into a beaker with a nitrogen stream. Several rinses of the cell with dry solvent followed by nitrogen flushing are recommended. NaCl solution cells are much more expensive than salt plates and must be stored in a desiccator.

Many chemists also acquire IR spectra of solids in KBr pellets, whose preparation and use will be left to other texts. Many of the foregoing methods are falling into disuse as IR spectrometers with universal attenuated total reflectance (ATR) accessories become more common. Here, the solid or liquid sample can be deposited directly onto a diamond window for spectral acquisition. As there is effectively no sample preparation with this method, there should be no barriers to the use of IR for compound characterization.

14.3   ULTRAVIOLET SPECTROSCOPY (UV)

Using volumetric techniques, a solution of precise molar concentration is made up in the desired spectroscopic grade solvent. Common solvents include diethyl ether, ethanol, hexane, and cyclohexane. The concentration is chosen such that the combination of the path length

Figure 14.3 A quartz UV cuvette with Teflon cap.

of the cuvette that will be used and an estimate of the extinction coefficient based on literature values will produce an absorption greater than 1. Calculation of the estimated absorption is based on the Beer-Lambert law (eq. 36), where c = concentration in M, d = path length in cm, and ε = extinction coefficient in $M^{-1}\ cm^{-1}$.

$$A = \varepsilon cd \qquad (36)$$

Cuvettes for UV absorption spectroscopy (Fig. 14.3) are made from quartz, which absorbs very little light in the 200 to 400 nm range. This is in contrast with conventional borosilicate glass (Pyrex), which absorbs significantly at $\lambda < 300$ nm. Generally, UV cuvettes come in pairs with matched optical properties. Quartz inserts are also available to reduce the path length of light through the solution from 1 to 0.1 cm or even 0.01 cm. When a compound has both strong and weak absorption bands, such inserts can permit the UV spectra of both bands to be obtained from a single solution. The solution is placed in a sample cuvette, and pure solvent is placed in a reference cuvette. The optical surfaces are wiped clean with tissue and the spectrum recorded. After solvent removal, the cuvettes are rinsed with pure solvent and flushed dry with nitrogen.

14.4 COMBUSTION ANALYSIS

If a compound is a solid, it should be recrystallized, preferably from a volatile solvent. It should then be pumped on overnight in the vial in which it will be sent for analysis. If it is a liquid, it can be purified by chromatography or distilled. Kugelrohr distillation is often used for this purpose—recall that it merely separates volatiles from non-volatiles, but it is often effective in removing the slight amount of color in a compound. If the compound has been purified by chromatography, is single spot on TLC, is pumped on overnight, and is pure, it will frequently hit. If it is a viscous liquid, removing all of the solvent may be more difficult. Heating with a heat gun may be required. A somewhat archaic piece of glassware to warm a sample under vacuum is a "drying pistol" (Fig. 14.4). Some compound

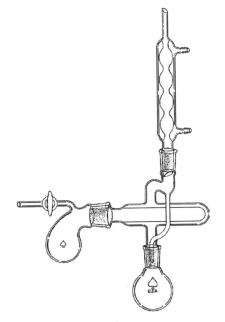

Figure 14.4 The Aberhalden drying pistol. A solvent with a bp below the mp of the solid to be dried is chosen. Its refluxing around the barrel of the vacuum chamber warms the solid to drive off any traces of solvent. If water is to be removed, a desiccant can be placed in the reservoir to the left (in the "handle").

classes, like nucleosides, are highly hygroscopic and analyze as hydrates. Recalculating the percentages of each element to include some water is acceptable with such difficult compounds. However, the only acceptable stoichiometries are 1 H_2O and $\frac{1}{2}H_2O$. Adding in fractional molecularities of water to get the combustion analysis to hit is manipulating the data. The amount of compound required for a combustion analysis is set by the policies of the analyst, but typically is 10 mg for carbon, hydrogen, and, if the compound includes it, nitrogen (CHN) in duplicate. More sample is required if other elements are to be determined.

14.5   MASS SPECTROMETRY

Mass spectral fragmentation patterns have received strong emphasis in older teachings of organic structure elucidation. However, these data are mostly available when using electron impact for ionization, and this ionization method is falling into disuse. More modern ionization methods are "soft," meaning that they do not lead to fragmentation. The molecular ion is therefore detected, which after all is the prime piece of information sought in a mass spectrum. Soft ionization methods include fast atom bombardment (FAB), matrix-assisted laser desorption-ionization (MALDI), electrospray ionization (ESI), and chemical ionization, particularly atmospheric pressure chemical ionization (APCI). These methods typically create charged adducts of analyte molecules with H^+, metal ions, or both. The loss of the information that was formerly available from fragmentation patterns is small, as information on molecular substructures is now more reliably determined from modern two-dimensional NMR techniques.

For the most part, obtaining a mass spectrum of a synthetic product has meant submitting a sample to a service facility. This is changing, though, as many facilities are taking advantage of "open access" LCMS instruments to permit chemists to obtain MS data themselves. Sample preparation protocols are unique to each type of instrument and facility, as is the software that analyzes the data. A capability often available in software, however, is matching of the molecular ion and the pattern of isotope peaks to possible molecular formulas.

14.6   CRYSTALLOGRAPHY

It is increasingly common for synthetic chemists to use X-ray crystal-lography to prove the structures of intermediates, and even to solve the structures themselves. Powerful computer software has put this task within the reach of many graduate students. However, growing of the crystals needed to obtain a structure is still a bit of an art, and determination of when a crystal is likely to give a good data set requires experience. If a crystallographer is nearby, they are a wealth of assistance and information. If not, available resources include this Web site: http://www.xray.ncsu.edu/GrowXtal.html.

CHAPTER

15

CLEANING UP AFTER THE REACTION

It is common to find ground glass joints that have become frozen or stuck during a reaction. This is typically the result of a reagent or solvent leaching away the grease. Approaches to freeing stuck joints are presented in Appendix 5.

When glassware has been used for reactions involving bases, and in some other instances like aluminum or other metal reagents, a fog may be seen on the glass that may be a metal hydroxide. These films hold onto the glass surface tightly, but rinsing the glassware with dil. HCl may remove them. For stubborn organic residues on glassware, including greases, chlorinated hydrocarbons often work best to remove them. When none of these tactics work to clean up glassware, a bath of chromic acid (or modern and more environmentally friendly cleaners like Nochromix®) is often considered. These methods may be very useful for volumetric apparatus in analytical chemistry; however, they are much less successful for organic residues.

Another choice is a cleaning bath formed from KOH in an alcohol solvent like isopropanol. This type of cleaning will likely be required by a glassblower before he will work on a piece of glassware that has been used for reactions. This treatment is very effective at removing all traces of silicon grease from joints and stopcocks, especially if warmed, but hydrocarbon grease (e.g., Apiezon) is impervious to a base bath. Glassware can be immersed for up to 30 minutes, but be certain not to leave glass in an alkali bath longer than needed to clean it. Prolonged immersion, even at ambient temperature, damages ground-glass joints, dissolves glass frits, and leaves glass surfaces etched. Remove the item using gloves or tongs, but use care because it will be slippery. Following cleaning in a base bath, a short soaking with 1 N HNO_3 will stop the attack of alkali on the glass.

The cleaning solution is prepared either by dissolving 100 g of KOH pellets in 50 mL of H_2O and, after cooling, making up to 1 L

The Synthetic Organic Chemist's Companion, by Michael C. Pirrung
Copyright © 2007 John Wiley & Sons, Inc.

with isopropanol or by adding 1 to 2 L of 95% ethanol to 120 mL of H_2O containing 120 g of NaOH or KOH. These baths have a relatively long lifetime, limited only by their effectiveness (in the judgment of the chemist). Only a steel container should be used for such an alkali bath. Glass could be easily shattered, spreading caustic solution everywhere, and of course will eventually be dissolved. Should the worst happen and the alkali/alcohol bath catch fire, plastic containers would melt and spread flaming caustic solution everywhere.

Syringes must be disassembled for cleaning, following which they may be dried in batches, cooled, and stored in a desiccator for future use. Drying is done in an oven (>130°C); usually 30 minutes are sufficient. Tuberculin syringes are numerically coded on the barrel and plunger and should be reassembled by matching the numbers. If one part is broken, most groups maintain a store of parts from which another plunger or barrel may be obtained. If the fit seems good, it is permissible to use syringes with mismatched numbers. If no fit is found, the unbroken part may be left in the store. Multi-fit syringes do not have numbers, so all fitting must be by trial and error.

Cleaning solutions must be pushed or pulled through most needles. Recall from the discussion of syringe use that reagent remains in the syringe needle following the transfer. This liquid must be removed and/or quenched. When using organometallics, aqueous tetrahydrofuran provides a useful quenching agent that will keep the generated metal hydroxide in solution. Mg and Al reagents may leave residues that require dil. HCl to remove. Sometimes needles come with cleaning wires that can be inserted to remove obstructions. The wire can be held with pliers. If needles (or syringes) become drastically clogged and/or stuck, an ultrasonic cleaner will often resurrect them.

CHAPTER

16

SPECIFIC EXAMPLES

This example is one step (boxed) of a multi-step sequence (eq. 37) to prepare a compound useful in synthetic methodology.

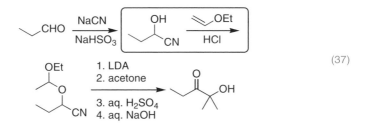

(37)

16.1 THE EXPERIMENTAL*

To 42.4 g (0.5 mol) of 2-hydroxybutyronitrile containing one drop of concentrated HCl was added 48 mL (36 g, 0.5 mol) of ethyl vinyl ether at a rate such that the temperature was maintained at ca. 50°C. After the addition was complete, the mixture was heated at 90°C for 2 h. Distillation directly from the reaction flask provided 65.7 g (84%) of 2-(1-ethoxyethoxy)-butyronitrile as a mixture of diastereomers, bp 84–96°C (18 torr).

16.2 THE *ORG. SYN.* PREP†

A 1 L, three-necked, round-bottomed flask is equipped with a condenser topped with a calcium chloride drying tube, a magnetic

---

* Reprinted with permission from Heathcock et al., 1980. Copyright 1980, American Chemical Society.
† Reprinted with permission from Young, Buse, and Heathcock (1985) by John Wiley & Sons, Inc.

stirring bar, a 500 mL pressure-equalizing addition funnel, and a thermometer. The flask is charged with 174 g (2.05 mol) of 2-hydroxy-butanenitrile to which 0.5 mL of concentrated hydrochloric acid has been added. The addition funnel is charged with 221 g (3.07 mol) of ethyl vinyl ether (Note 5), which is then added dropwise to the stirring cyanohydrin at such a rate that the temperature is maintained at ca. 50°C. When the addition is complete, the mixture is heated to 90°C for 4 h. The condenser is replaced with a distillation head and the dropping funnel and thermometer are replaced with stoppers. Direct distillation of the gold-yellow solution from the reaction flask yields 226–277 g (70–86%) of nearly pure 2-[(1'-ethoxy)-1-ethoxy]butanenitrile, bp 85–84°C (30 mm), as a colorless liquid (Note 6).

Notes

5. Ethyl vinyl ether was obtained from Aldrich Chemical Company and was used without further purification.
6. The IR spectrum (neat) shows absorption at 2970, 1425, and 1385 cm^{-1}. The C$\equiv$N absorption is not observed.

16.3   COMPARISON

An obvious difference between these two descriptions of the same transformation is that the *Org. Syn.* prep describes the apparatus in some detail. It also is conducted on a much larger scale. Even the experimental prep is far beyond the 25 mmol scale that has been the focus of this book.

CHAPTER

17

STRATEGIES FOR REACTION OPTIMIZATION

When conducting a reaction from the literature for the first time, the outcome (yield, purity, time, convenience) is unlikely to be as good as reported, or even as good as will likely be the case once the chemist has experience with it. Because the written detail available on a reaction is far exceeded by the amount of expertise and chemical know-how that goes into conducting it, this is to be expected. The chemists reporting the reaction presumably did a significant amount of optimization and provided their "best" procedure. The need for extensive experimental variation to meet or exceed the published results is not expected. However, there are also variations in the source and quality of reagents, available apparatus, and even the climate that can affect reaction outcomes. We experienced this in our lab in the preparation of the very moisture-sensitive nucleoside phosphoramidites (Fig. 13.4), the building blocks of DNA synthesis. This chemistry was developed in the high altitudes and low humidity of Boulder, Colorado, but when conducted in the summer in muggy Durham, North Carolina, reactions often gave only half the reported yields. Although optimization of conditions for known reactions is not necessary, a fresh perspective or different set of chemical experiences or skills can always be brought to bear to improve upon earlier reports. Such contributions should be welcomed by the community of synthetic chemists and are very worthy of publication.

Optimization efforts take on much greater importance when developing new reactions. A cliché of the scientific method is "change only one variable at a time." While good advice to the novice scientist (e.g., one in grade school) involved in deductive investigation, it is a dictum that is widely over-applied. It is much less useful for the inductive investigations of synthetic organic chemistry. The reasons behind this are well-known and readily illustrated with a three-dimensional plot (Fig. 17.1) of the outcome of an experiment (A) versus two variables, calcium chloride and magnesium chloride concentration. Assume that the initial experiment was conducted with 150 mM $CaCl_2$, giving A = 0.2 (the point nearest the viewer). Changing only one variable at a time means being restricted to movement along the axes of this plot. Adding increasing concentrations of $MgCl_2$ causes a decrease of A, so the experimenter would (erroneously) conclude that $MgCl_2$ should not be present. Decreasing the concentration of $CaCl_2$ causes an increase of A, until its concentration is very low, when A drops. The conclusion of these experiments, following the one variable at a time dictum, is that the optimum experimental conditions are low $CaCl_2$ and no $MgCl_2$. The appearance of the complete experimental response surface shows that the optimum outcome is with low $MgCl_2$ and no $CaCl_2$, however.

Statistical methods to design and interpret experiments involving multiple variables have been available for some time. They have likely not been applied in synthetic chemistry as broadly as they might be, though texts describing such work are available (Carlson and Carlson, 2005). There are several reasons for this. One is the

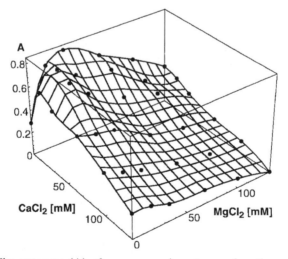

Figure 17.1 The outcome (A) of many experiments as a function of the ion concentration. Reprinted from Zauner, K.P.; Conrad, M. (2000). Enzymatic pattern processing. *Naturwissenschaften* 87, 360–362, Figure 1. Copyright 2000, Springer. With kind permission of Springer Science and Business Media.

dogma that only one variable should be changed at a time. Another is the difficulty that nonspecialists have had in using the computer programs that perform multivariable optimizations. Another is the number of experiments that may be required to adequately map out the experimental response surface (note the number of data points in Fig. 17.1). However, the relatively recent interest in methods for high-throughput and automated experimentation has made collection of these data much less burdensome. Yet it might also be considered that the capability to perform many experiments easily enables optimization to be performed empirically, without the need for statistical design. At the very least chemists could avoid the pitfall described in the example by selecting a number of values for each variable and examining a reasonable number of the combinations that differ from the initial conditions.

Another limitation of statistical design methods is that they apply only to continuous variables, such as concentration and temperature. Experienced synthetic chemists know that nonnumerical variables such as solvent, order of addition, and workup methods can profoundly affect reaction outcome. The ability to conduct many trials of a reaction is often the only means to be certain that the best procedure has been found. In the industrial setting, hardware permitting high-throughput experimentation may be available, but of course the chemist still needs to tell it what experiments to perform. The laboratory experience gained from optimizing synthetic reactions is indispensable to this decision-making.

Strategies to perform many trials of a reaction are not complex. Reaction scale can be kept as small as possible to minimize starting material consumption. At the lower limit (0.5 mg or less), reactions can be designed only for analysis of the outcome by TLC or GC, with no intention of isolating the product. The chemist is looking merely for a single product with properties consistent with the expected product. Racks of vials can be used as reaction vessels for a dozen or more reactions. In this sort of broad screening, the reactions need not be the same. For example, there are hundreds of reagents to oxidize secondary alcohols to ketones. Replicate vials with the reaction substrate could be set up and a different oxidant added to each. These sorts of approaches are essential to multi-step, target-directed syntheses, where precious intermediates at the "frontier" of the total synthesis have been hard won through weeks or even months of effort.

APPENDIX

1

NMR SPECTRAL DATA OF COMMON CONTAMINANTS OF ORGANIC REACTION PRODUCTS

The Synthetic Organic Chemist's Companion, by Michael C. Pirrung
Copyright © 2007 John Wiley & Sons, Inc.

Table A1.1 ^{1}H NMR Data of Common Solvents and Contaminants

	proton	mult	CDCl₃	(CD₃)₂CO	(CD₃)₂SO	C₆D₆	CD₃CN	CD₃OD	D₂O
solvent residual peak			7.26	2.05	2.50	7.16	1.94	3.31	4.79
H₂O		s	1.56	2.84	3.33	0.40	2.13	4.87	
acetic acid	CH₃	s	2.10	1.96	1.91	1.55	1.96	1.99	2.08
acetone	CH₃	s	2.17	2.09	2.09	1.55	2.08	2.15	2.22
acetonitrile	CH₃	s	2.10	2.05	2.07	1.55	1.96	2.03	2.06
benzene	CH	s	7.36	7.36	7.37	7.15	7.37	7.33	
tert-butyl alcohol	CH₃	s	1.28	1.18	1.11	1.05	1.16	1.40	1.24
	OH	s			4.19	1.55	2.18		
tert-butyl methyl ether	CCH₃	s	1.19	1.13	1.11	1.07	1.14	1.15	1.21
	OCH₃	s	3.22	3.13	3.08	3.04	3.13	3.20	3.22
chloroform	CH	s	7.26	8.02	8.32	6.15	7.58	7.90	
cyclohexane	CH₂	s	1.43	1.43	1.40	1.40	1.44	1.45	
1,2-dichloroethane	CH₂	s	3.73	3.87	3.90	2.90	3.81	3.78	
dichloromethane	CH₂	s	5.30	5.63	5.76	4.27	5.44	5.49	
diethyl ether	CH₃	t,7	1.21	1.11	1.09	1.11	1.12	1.18	1.17
	CH₂	q,7	3.48	3.41	3.38	3.26	3.42	3.49	3.56
diglyme	CH₂	m	3.65	3.56	3.51	3.46	3.53	3.61	3.67
	CH₂	m	3.57	3.47	3.38	3.34	3.45	3.58	3.61
	OCH₃	s	3.39	3.28	3.24	3.11	3.29	3.35	3.37
1,2-dimethoxyethane	CH₃	s	3.40	3.28	3.24	3.12	3.28	3.35	3.37
	CH₂	s	3.55	3.46	3.43	3.33	3.45	3.52	3.60

dimethylacetamide	CH_3CO	s	2.09	1.97	1.96	1.60	1.97	2.07	2.08
	NCH_3	s	3.02	3.00	2.94	2.57	2.96	3.31	3.06
	NCH_3	s	2.94	2.83	2.78	2.05	2.83	2.92	2.90
dimethylformamide	CH	s	8.02	7.96	7.95	7.63	7.92	7.97	7.92
	CH_3	s	2.96	2.94	2.89	2.36	2.89	2.99	3.01
	CH_3	s	2.88	2.78	2.73	1.86	2.77	2.86	2.85
dimethyl sulfoxide	CH_3	s	2.62	2.52	2.54	1.68	2.50	2.65	2.71
dioxane	CH_2	s	3.71	3.59	3.57	3.35	3.60	3.66	3.75
ethanol	CH_3	t,7	1.25	1.12	1.06	0.96	1.12	1.19	1.17
	CH_2	q,7	3.72	3.57	3.44	3.34	3.54	3.60	3.65
	OH	s	1.32	3.39	4.63		2.47		
ethyl acetate	CH_3CO	s	2.05	1.97	1.99	1.65	1.97	2.01	2.07
	CH_2CH_3	q,7	4.12	4.05	4.03	3.89	4.06	4.09	4.14
	CH_2CH_3	t,7	1.26	1.20	1.17	0.92	1.20	1.24	1.24
ethyl methyl ketone	CH_3CO	s	2.14	2.07	2.07	1.58	2.06	2.12	2.19
	CH_2CH_3	q,7	2.46	2.45	2.43	1.81	2.43	2.50	3.18
	CH_2CH_3	t,7	1.06	0.96	0.91	0.85	0.96	1.01	1.26
ethylene glycol	CH	s	3.76	3.28	3.34	3.41	3.51	3.59	3.65
"grease"	CH_3	m	0.86	0.87		0.92	0.86	0.88	
	CH_2	br s	1.26	1.29		1.36	1.27	1.29	
n-hexane	CH_3	t	0.88	0.88	0.86	0.89	0.89	0.90	
	CH_2	m	1.26	1.28	1.25	1.24	1.28	1.29	
HMPA	CH_3	d,9.5	2.65	2.59	2.53	2.40	2.57	2.64	2.61
methanol	CH_3	s	3.49	3.31	3.16	3.07	3.28	3.34	3.34
	OH	s	1.09	3.12	4.01		2.16		

Table A1.1 (*Continued*)

proton	mult	CDCl$_3$	(CD$_3$)$_2$CO	(CD$_3$)$_2$SO	C$_6$D$_6$	CD$_3$CN	CD$_3$OD	D$_2$O
nitromethane CH$_3$	s	4.33	4.43	4.42	2.94	4.31	4.34	4.40
n-pentane CH$_3$	t,7	0.88	0.88	0.86	0.87	0.89	0.90	
CH$_2$	m	1.27	1.27	1.27	1.23	1.29	1.29	
2-propanol CH$_3$	d,6	1.22	1.10	1.04	0.95	1.09	1.50	1.17
CH	sep,6	4.04	3.90	3.78	3.67	3.87	3.92	4.02
pyridine CH(2)	m	8.62	8.58	8.58	8.53	8.57	8.53	8.52
CH(3)	m	7.29	7.35	7.39	6.66	7.33	7.44	7.45
CH(4)	m	7.68	7.76	7.79	6.98	7.73	7.85	7.87
silicone grease CH$_3$	s	0.07	0.13		0.29	0.08	0.10	
tetrahydrofuran CH$_2$	m	1.85	1.79	1.76	1.40	1.80	1.87	1.88
CH$_2$O	m	3.76	3.63	3.60	3.57	3.64	3.71	3.74
toluene CH$_3$	s	2.36	2.32	2.30	2.11	2.33	2.32	
CH(o/p)	m	7.17	7.1–7.2	7.18	7.02	7.1–7.3	7.16	
CH(m)	m	7.25	7.1–7.2	7.25	7.13	7.1–7.3	7.16	

Reprinted with Permission from Gottlieb, Kotlyar, and Nudelman, 1997. Copyright 1997, American Chemical Society.

Table A1.2 ^{13}C NMR Data of Common Solvents and Contaminants

		CDCl$_3$	(CD$_3$)$_2$CO	(CD$_3$)$_2$SO	C$_6$D$_6$	CD$_3$CN	CD$_3$OD	D$_2$O
solvent signals		77.16 ± 0.06	29.84 ± 0.01 206.26 ± 0.13	39.52 ± 0.06	128.06 ± 0.02	1.32 ± 0.02 118.26 ± 0.02	49.00 ± 0.01	
acetic acid	CO	175.99	172.31	171.93	175.82	173.21	175.11	177.21
	CH$_3$	20.81	20.51	20.95	20.37	20.73	20.56	21.03
acetone	CO	207.07	205.87	206.31	204.43	207.43	209.67	215.94
	CH$_3$	30.92	30.60	30.56	30.14	30.91	30.67	30.89
acetonitrile	CN	116.43	117.60	117.91	116.02	118.26	118.06	119.68
	CH$_3$	1.89	1.12	1.03	0.20	1.79	0.85	1.47
benzene	CH	128.37	129.15	128.30	128.62	129.32	129.34	
tert-butyl alcohol	C	69.15	68.13	66.88	68.19	68.74	69.40	70.36
	CH$_3$	31.25	30.72	30.38	30.47	30.68	30.91	30.29
tert-butyl methyl	OCH$_3$	49.45	49.35	48.70	49.19	49.52	49.66	49.37
ether	C	72.87	72.81	72.04	72.40	73.17	74.32	75.62
	CCH$_3$	26.99	27.24	26.79	27.09	27.28	27.22	26.60
chloroform	CH	77.36	79.19	79.16	77.79	79.17	79.44	
cyclohexane	CH$_2$	26.94	27.51	26.33	27.23	27.63	27.96	
1,2-dichloroethane	CH$_2$	43.50	45.25	45.02	43.59	45.54	45.11	
dichloromethane	CH$_2$	53.52	54.95	54.84	53.46	55.32	54.78	
diethyl ether	CH$_3$	15.20	15.78	15.12	15.46	15.63	15.46	14.77
	CH$_2$	65.91	66.12	62.05	65.94	66.32	66.88	66.42
diglyme	CH$_3$	59.01	58.77	57.98	58.66	58.90	59.06	58.67
	CH$_2$	70.51	71.03	69.54	70.87	70.99	71.33	70.05
	CH$_2$	71.90	72.63	71.25	72.35	72.63	72.92	71.63

Table A1.2 (Continued)

		CDCl₃	(CD₃)₂CO	(CD₃)₂SO	C₆D₆	CD₃CN	CD₃OD	D₂O
1,2-dimethoxyethane	CH₃	59.08	58.45	58.01	58.68	58.89	59.06	58.67
	CH₂	71.84	72.47	17.07	72.21	72.47	72.72	71.49
dimethylacetamide	CH₃	21.53	21.51	21.29	21.16	21.76	21.32	21.09
	CO	171.07	170.61	169.54	169.95	171.31	173.32	174.57
	NCH₃	35.28	34.89	37.38	34.67	35.17	35.50	35.03
	NCH₃	38.13	37.92	34.42	37.03	38.26	38.43	38.76
dimethylformamide	CH	162.62	162.79	162.29	162.13	163.31	164.73	165.53
	CH₃	36.50	36.15	35.73	35.25	36.57	36.89	37.54
	CH₃	31.45	31.03	30.73	30.72	31.32	31.61	32.03
dimethyl sulfoxide	CH₃	40.76	41.23	40.45	40.03	41.31	40.45	39.39
dioxane	CH₂	67.14	67.60	66.36	67.16	67.72	68.11	67.19
ethanol	CH₃	18.41	18.89	18.51	18.72	18.80	18.40	17.47
	CH₂	58.28	57.72	56.07	57.86	57.96	58.26	58.05
ethyl acetate	CH₃CO	21.04	20.83	20.68	20.56	21.16	20.88	21.15
	CO	171.36	170.96	170.31	170.44	171.68	172.89	175.26
	CH₂	60.49	60.56	59.74	60.21	60.98	61.50	62.32
	CH₃	14.19	14.50	14.40	14.19	14.54	14.49	13.92
ethyl methyl ketone	CH₃CO	29.49	29.30	29.26	28.56	29.60	29.39	29.49
	CO	209.56	208.30	208.72	206.55	209.88	212.16	218.43
	CH₂CH₃	36.89	36.75	35.83	36.36	37.09	37.34	37.27
	CH₂CH₃	7.86	8.03	7.61	7.91	8.14	8.09	7.87
ethylene glycol	CH₂	63.79	64.26	62.76	64.34	64.22	64.30	63.17
"grease"	CH₂	29.76	30.73	29.20	30.21	30.86	31.29	

n-hexane	CH$_3$	14.14	14.34	13.88	14.32	14.43	14.45	
	CH$_2$(2)	22.70	23.28	22.05	23.04	23.40	23.68	
	CH$_2$(3)	31.64	32.30	30.95	31.96	32.36	32.73	
HMPA	CH$_3$	36.87	37.04	36.42	36.88	37.10	37.00	36.46
methanol	CH$_3$	50.41	49.77	48.59	49.97	49.90	49.86	49.50
nitromethane	CH$_3$	62.50	63.21	63.28	61.16	63.66	63.08	63.22
n-pentane	CH$_3$	14.08	14.29	13.28	14.25	14.37	14.39	
	CH$_2$(2)	22.38	22.98	21.70	22.72	23.08	23.38	
	CH$_2$(3)	34.16	34.83	33.48	34.45	34.89	35.30	
2-propanol	CH$_3$	25.14	25.67	25.43	25.18	25.55	25.27	24.38
	CH	64.50	63.85	64.92	64.23	64.30	64.71	64.88
pyridine	CH(2)	149.90	150.67	149.58	150.27	150.76	150.07	149.18
	CH(3)	123.75	124.57	123.84	123.58	127.76	125.53	125.12
	CH(4)	135.96	136.56	136.05	135.28	136.89	138.35	138.27
silicone grease	CH$_3$	1.04	1.40		1.38		2.10	
tetrahydrofuran	CH$_2$	25.62	26.15	25.14	25.72	26.27	26.48	25.67
	CH$_2$O	67.97	68.07	67.03	67.80	68.33	68.83	68.68
toluene	CH$_3$	21.46	21.46	20.99	21.10	21.50	21.50	
	C(i)	137.89	138.48	137.35	137.91	138.90	138.85	
	CH(o)	129.07	129.76	128.88	129.33	129.94	129.91	
	CH(m)	128.26	129.03	128.18	128.56	129.23	129.20	
	CH(p)	125.33	126.12	125.29	125.68	126.28	126.29	

APPENDIX

2

SYNTHETIC SOLVENT SELECTION CHART

The chart on the following page (Fig. A2.1, as well as on the foldout from the inside back cover) is based on principal components analysis of the properties mp, bp, dielectric constant (ε); dipole moment (μ); refractive index (η); E_T (spectroscopically determined effect on a solvatochromatic dye); and *log P* for >80 solvents (Carlson, Lunstedt, and Albano, 1985). The eigenvectors were projected into two dimensions t_1 and t_2, representing a conflation of properties that most differentiate these solvents, accounting for around 80% of their variance. Polarity correlates with t_1, while polarizability correlates with t_2. Solvents in the vicinity of one another in this chart should have similar properties.

The Synthetic Organic Chemist's Companion, by Michael C. Pirrung
Copyright © 2007 John Wiley & Sons, Inc.

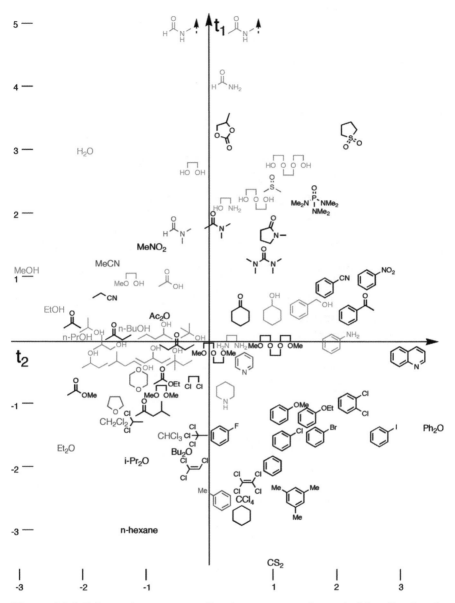

Figure A2.1 Solvents known to swell polystyrene are shown in blue. Protic solvents are shown in red. Aprotic solvents are shown in black. *N*-methylformamide and *N*-methylacetamide are shown slightly lower than their t_1 coordinates, 5.4 and 5.3, to compact the chart.

APPENDIX

3

RECIPES FOR TLC STAINS

Staining solutions are ideally stored in 100 mL wide-mouth jars covered with aluminum foil, a watch glass, or a screw cap to minimize evaporation.

A3.1 GENERAL STAINS

Phosphomolybdic Acid (PMA): Dissolve 10 g of phosphomolybdic acid in 100 mL of absolute ethanol.

Potassium Permanganate: Dissolve 1.5 g of $KMnO_4$, 10 g of K_2CO_3, and 1.25 mL of 10% NaOH in 200 mL of water.

Vanillin: Dissolve 15 g of vanillin in 250 mL of ethanol and add 2.5 mL of conc. sulfuric acid.

p-Anisaldehyde A: To 135 mL of absolute ethanol add 5 mL of conc. sulfuric acid and 1.5 mL of glacial acetic acid. Allow the solution to cool to room temperature. Add 3.7 mL of p-anisaldehyde. Stir the solution vigorously to ensure homogeneity. Store refrigerated.

p-Anisaldehyde B: Prepare a solution of anisaldehyde : $HClO_4$: acetone : water in the proportions 1 : 10 : 20 : 80.

p-Anisaldehyde C: Dissolve 2.5 mL of conc. sulfuric acid and 15 mL of p-anisaldehyde in 250 mL of 95% ethanol.

Cerium Molybdate (Hanessian's Stain): To 235 mL of distilled water add 12 g of ammonium molybdate, 0.5 g of ceric ammonium molybdate, and 15 mL of conc. sulfuric acid. Wrap the jar with aluminum foil as the stain may be somewhat photo-sensitive.

The Synthetic Organic Chemist's Companion, by Michael C. Pirrung
Copyright © 2007 John Wiley & Sons, Inc.

A3.2   SPECIALIZED STAINS

Dinitrophenylhydrazine (DNP): Dissolve 12 g of 2,4-dinitro-phenylhydrazine, 60 mL of conc. sulfuric acid, and 80 mL of water in 200 mL of 95% ethanol.

Ninhydrin: Dissolve 0.3 to 1.5 g of ninhydrin in 100 mL of *n*-butanol. Then add 3.0 mL of acetic acid.

Bromocresol Green: Dissolve 0.04 g of bromocresol green in 100 mL of absolute ethanol, which should give a colorless solution. Add a 0.1 M solution of aqueous NaOH dropwise until a blue color just appears.

Ceric Ammonium Sulfate: Prepare a 1% (w/v) solution of cerium (IV) ammonium sulfate in 50% phosphoric acid.

Ceric Ammonium Molybdate: Dissolve 0.5 g of ceric ammonium sulfate $(Ce(NH_4)_4(SO_4)_4 \cdot 2H_2O)$ and 12 g of ammonium molybdate $(NH_4)_6Mo_7O_{24} \cdot 4H_2O)$ in 235 mL of water. Add 15 mL of conc. sulfuric acid.

Cerium Sulfate: Prepare a 10% (w/v) solution of cerium (IV) sulfate in 15% sulfuric acid.

Ferric Chloride: Prepare a 1% (w/v) solution of 1% ferric (III) chloride in 50% aqueous methanol.

Ehrlich's Reagent: Dissolve 1.0 g of *p*-dimethylaminobenzalde-hyde in 75 mL of MeOH and add 50 mL of conc. HCl.

Dragendorff-Munier Stain: Dissolve 10 g of KI in 40 mL water. Add 1.5 g of $Bi(NO_3)_3$ and 20 g of tartaric acid in 80 mL of water. Stir for 15 minutes and then filter.

APPENDIX

4

MIXOTROPIC SERIES

Table A3

Water***
Acetonitrile
Methanol***
Acetic acid
Ethanol
Isopropanol
Acetone
n-Propanol
Dioxane
Tetrahydrofuran
tert-Butanol
sec-Butanol
Butanone
Cyclohexanone
n-Butanol
Cyclohexanol
Ethyl acetate
Pyridine***
Ethyl formate
Diethyl ether
Diethoxymethane
Butyl acetate
Nitromethane
Dichloromethane
Chloroform***
Benzene

The Synthetic Organic Chemist's Companion, by Michael C. Pirrung
Copyright © 2007 John Wiley & Sons, Inc.

Table A3 (*Continued*)

Trichloroethylene
Toluene
Xylenes
Cyclopentane
Cyclohexane
n-Hexane
n-Heptane***
Petroleum ether

***Solvents representative of one of the five classes of solvents based on the strength of intermolecular interactions.

APPENDIX

5

STUCK JOINTS

Frozen ground glass joints are a common problem. To keep joints from freezing, grease the joint every time it is assembled. Nevertheless, even with good greasing, joints can still seize. Grease may be leached out by the solvent, especially if it is heated. Solids stuck in the joint (e.g., salts from a reaction) can also cause a joint to seize. Several strategies to free stuck joints are available, but creativity is typically also essential to solve such situations.

SAFETY NOTE

Trying to free stuck glass joints can result in serious bodily injury or death if the glass breaks in the process. Wear eye protection and sturdy work gloves and wrap the glass in a heavy lab towel to lower the potential for serious cuts or injury. Do not use heat or flame if there is any potential for hazardous or flammable residues or vapors to ignite. Never heat a closed system. If you are not comfortable with assessing or understanding the risk involved, do not attempt freeing joints yourself.

The following methods can be conducted once the flask has been emptied (through a sidearm or another opening), cleaned, and dried to ensure that there are no harmful residues or solvents present:

1. Soak the joint overnight, either in a base bath (KOH/ isopropanol) or an ultrasonicator bath. Look for uniform wetting of the ground joint surfaces, which indicates liquid penetration.
2. Put a solvent, glycerol, or a penetrating oil on the outside of the joint and try to work it down into the joint as the parts are rocked. Again, look for a "clear" joint.

The Synthetic Organic Chemist's Companion, by Michael C. Pirrung
Copyright © 2007 John Wiley & Sons, Inc.

3. Apply heat to the outer glass joint so that it expands, freeing it from the inner joint. For this method to work well, it is essential that the outer joint expands rapidly without similar expansion of the inner joint. Because their heating rates are too slow to promote differential expansion, heat guns or Bunsen burners typically do not work well in this method. Heat the joint with a glassblowing torch in a cool to moderate (yellow) flame with just a touch of oxygen. Direct the flame to any adhesion points that are visible, but do not hold the flame on any one spot for too long—you will lose the effect of differential expansion. After a few seconds of rotation of the joint in the flame, use an insulated, gloved hand or a hook to try to separate the parts. Glassware must be annealed if a torch has been used on the joint.

4. Put the apparatus through an annealing cycle in the glass shop's annealing oven.

If a glass stopper has become stuck in a joint, hold the head of the stopper with a gloved hand or a clamp. Place a wooden stick or dowel against the outer joint and tap lightly but sharply on it with a hammer. Assistance to hold one of the items or to catch the parts as they separate can be helpful with this technique.

If this does not work, put a lab towel over the head of the stopper and loosely place a crescent (adjustable) wrench on the head. Do not let the metal touch the head. Try to rotate the stopper with the wrench, but do not apply so much force that you break the stopper (which, unfortunately, is easy to do). If the stopper breaks and it is hollow, smash out its remains. If it is solid, move on to other options.

When a separatory funnel with Teflon stopcock is used, the stopcock should not be overtightened because it will warp, and should not be exposed to sudden heat, since Teflon expands much faster than glass. Conversely, stuck Teflon stopcocks can be freed by immersing them in dry ice/acetone.

If none of these methods work, ask a glassblower, who may have access to specialized tools like glass joint or stopper pullers (similar to gear pullers, for you mechanics).

APPENDIX

6

ACIDITIES OF ORGANIC FUNCTIONAL GROUPS

These acidities are expressed as pK_a compared to water. Those in Figure A6.1 were determined in aqueous media, and those on the second page were determined by extrapolation from aqueous media. The hydrogen to which the pK_a refers is explicitly shown. These data came from several sources (House, 1972; March, 1992; Reichardt, 2002).

The Synthetic Organic Chemist's Companion, by Michael C. Pirrung
Copyright © 2007 John Wiley & Sons, Inc.

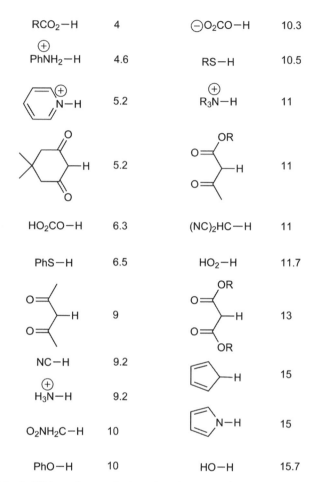

RCO$_2$—H	4	$^{\ominus}$O$_2$CO—H	10.3
PhNH$_2$—H (+)	4.6	RS—H	10.5
N—H (+)	5.2	R$_3$N—H (+)	11
—H	5.2	—H	11
HO$_2$CO—H	6.3	(NC)$_2$HC—H	11
PhS—H	6.5	HO$_2$—H	11.7
—H	9	—H	13
NC—H	9.2	—H	15
H$_3$N—H (+)	9.2	N—H	15
O$_2$NH$_2$C—H	10		
PhO—H	10	HO—H	15.7

Figure A6.1 Acidities of organic functional groups (as expressed in pK_a) compared to water.

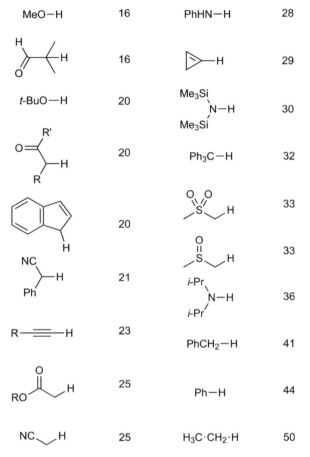

MeO—H	16	PhHN—H	28
	16		29
t-BuO—H	20		30
	20	Ph₃C—H	32
	20		33
			33
	21		36
R—≡—H	23	PhCH₂—H	41
	25	Ph—H	44
	25	H₃C·CH₂·H	50

Figure A6.1 Continued

APPENDIX

7

ACIDITIES OF ORGANIC FUNCTIONAL GROUPS IN DMSO

Acidities (expressed as pK_a) of organic functional groups in DMSO (Figure A7.1), (Bordwell, 1988). The hydrogen to which the pK_a refers is explicitly shown.

The Synthetic Organic Chemist's Companion, by Michael C. Pirrung
Copyright © 2007 John Wiley & Sons, Inc.

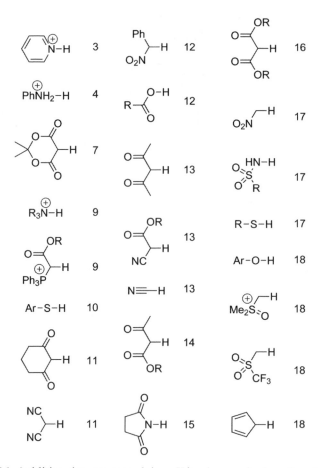

Figure A7.1 Acidities (as expressed in pK_a) of organic functional groups in DMSO.

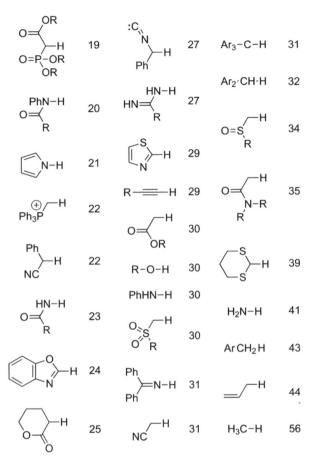

Figure A7.1 Continued

APPENDIX

8 WEB RESOURCES FOR SYNTHETIC CHEMISTRY

The dynamic nature of the World Wide Web makes the Web sites that are not maintained by mainline publishers impermanent. I have therefore elected to include in this appendix rather than in the main text Web sites that are maintained by individuals.

Alison Frontier—*Not Voodoo*
http://www.chem.rochester.edu/~nvd/
"Demystifying Synthetic Organic Laboratory Technique"

Hans Reich—*Organic Chemistry Info*
http://www.chem.wisc.edu/areas/organic/index-chem.htm

Al Garofalo—*Al's Notebook*
http://www.alsnotebook.com
"A collection of commonly used experimental procedures and other interesting stuff for synthetic chemists."

Thomas Sander—*WebReactions*
http://www.webreactions.net
"a state-of-the-art chemical reaction search solution. . . . The core of the system is the world's fastest reaction search engine based on the reaction classification system of J. B. Hendrickson".

The Synthetic Organic Chemist's Companion, by Michael C. Pirrung

BIBLIOGRAPHY

Armarego, W.L.F.; Perrin, D.D. (2000). *Purification of Laboratory Chemicals*, 4th ed. Boston: Butterworth-Heinemann.

Black, T.H. (1983). The preparation and reactions of diazomethane. *Aldrichimica Acta* **16**, 3–10.

Bordwell, F.G. (1988). Equilibrium acidities in dimethyl sulfoxide solution. *Acc. Chem. Res.* **21**, 456–463.

Caddick, S.; Aboutayab, K.; West, R. (1993). An intramolecular radical cyclisation approach to fused [1,2-a]indoles. *Synlett* 231–232.

Carlson, R.; Lunstedt, T.; Albano, C. (1985). Screening of suitable solvents in organic synthesis. Strategies for solvent selection. *Acta Chem. Scand. B* **39**, 79–91.

Carlson, R.; Carlson, J. (2005). *Design and Optimization in Organic Synthesis, 24,* 2nd ed. New York: Elsevier.

Committee on Prudent Practices for Handling, Storage, and Disposal of Chemicals in Laboratories, National Research Council. (1995). *Prudent Practices in the Laboratory: Handling and Disposal of Chemicals.* Washington, DC: National Academy Press.

Furniss, B.S.; Hannaford, A.J.; Smith, P.W.G.; Tatchell, A.R. (1989). *Vogel's Textbook of Practical Organic Chemistry,* 5th ed. New York: Longman.

Heathcock, C.H.; Buse, C.T.; Kleschick, W.A.; Pirrung, M.C.; Sohn, J.E.; Lampe, J. (1980). Acyclic stereoselection. 7. Stereoselective synthesis of 2-alkyl-3-hydroxy carbonyl compounds by aldol condensation. *J. Org. Chem.* **45**, 1066–1081.

House, H.O. (1972). *Modern Synthetic Reactions,* 2nd ed. Menlo Park: Benjamin.

Hoye, T.R.; Eklov, B.M.; Ryba, T.D.; Voloshin, M.; Yao, L.J. (2004). No-D NMR (no-deuterium proton NMR) spectroscopy: A simple yet powerful method for analyzing reaction and reagent solutions. *Org. Lett.* **6**, 953–956.

The Synthetic Organic Chemist's Companion, by Michael C. Pirrung
Copyright © 2007 John Wiley & Sons, Inc.

Kappe, C.O.; Stadler, A. (2005). *Methods and Principles in Medicinal Chemistry: Microwaves in Organic and Medicinal Chemistry.* Weinheim: Wiley-VCH.

March, J. (1992). *Advanced Organic Chemistry: Reactions, Mechanisms, and Structure*, 4th ed. New York: Wiley.

Pangborn, A. B.; Giardello, M. A.; Grubbs, R. H.; Rosen, R. K.; Timmers, F. J. (1996). Safe and convenient procedure for solvent purification. *Organometallics* **15**, 1518–1520.

Pirrung, M. C.; Deng, L.; Li, Z.; Park, K. (2002). Synthesis of 2,5-dihydroxy-3-(indol-3-yl)-benzoquinones by acid-catalyzed condensation. *J. Org. Chem.* **67**, 8374–8388.

Ratnayake, W.M.N. (2004). Overview of methods for the determination of trans fatty acids by gas chromatography, silver-ion thin-layer chromatography, silver-ion liquid chromatography, and gas chromatography/ mass spectrometry. *J. AOAC Int.* **87**, 523–539.

Reichardt, C. (2002). *Solvents and Solvent Effects in Organic Chemistry*, 3rd ed. Weinheim: Wiley-VCH.

Shanley, E.S.; Greenspan, F.P. (1947). Highly concentrated hydrogen peroxide. *Ind. Eng. Chem.* **39**, 1536–1543.

Shriver, D.F.; Drezdzon, M.A. (1986). *The Manipulation of Air-Sensitive Compounds,* 2nd ed. New York: Wiley.

Still, W. C.; Kahn, M.; Mitra, A. (1978). Rapid chromatographic technique for preparative separations with moderate resolution. *J. Org. Chem.* **43**, 2923–2925.

Williams, C.M.; Mander, L.N. (2001). Chromatography with silver nitrate. *Tetrahedron* **57**, 425–447.

Yamamoto, Y.; Nishimura, K.; Kiriyama, N. (1976). Studies on the metabolic products of *Aspergillus terreus.* I. Metabolites of the strain IFO 6123. *Chem. Pharm. Bull.* **24**, 1853–1859.

Young, S.D.; Buse, C.T.; Heathcock, C.H. (1985). 2-Methyl-2-(trimethylsiloxy)pentan-3-one. [3-Pentanone, 2-methyl-2-[(trimethylsilyl) oxy]-]. *Org. Synth.* **63**, 79–85.

Zauner, K.P.; Conrad, M. (2000). Enzymatic pattern processing. *Naturwissenschaften* **87**, 360–362.

Zubrick, J.W. (2003). *The Organic Chem Lab Survival Manual: A Student's Guide to Techniques,* 6th ed. New York: Wiley.

INDEX

Acetaldehyde, 54
Acetic anhydride, 57
Acetone, 50, 57, 162, 165
  coolant, 36, 131, 176
  deuterated, 105, 141
  reactant, 108
  solvent, 17, 132, 173
  wash, 77, 142, 147
Acetonitrile, 50, 57, 70, 162, 165,
    170
  deuterated, 105
  solvent, 17, 58, 97, 104, 111, 138,
    173
Acetylenes, 18
Acid
  acetic, 162, 165, 173
    as acid, 26
    by-product, 80, 81, 108
    reagent, 7
    solvent, 53, 85, 97, 108, 171,
      170
    quench, 66
  chromic, 143, 153
  diphenylacetic, 13, 14
  fluoboric, 108
  formic, 97, 144
  hydrazoic, 18
  hydrofluoric, 25, 108
  lactic, 108
  oxalic, 136
  peracetic, 7
  phosphomolybdic, 171
  phosphoric, 25, 172

  sulfuric, 171, 172
  tartaric, 172
  trifluoroperoxyacetic, 90
Acidities, 30, 72, 73, 177–183
Ac_2O, 57
Addition
  inverse, 78
  normal, 78
Adsorption, 113
$AgNO_3$, see Silver nitrate
Aldol condensation, 70, 132
Alkene, 100
Alumina, 58, 94, 97, 113, 115, 138
Aluminum foil, 8, 17, 80, 86, 139,
    171
Ammonia
  liquid, 27, 41, 60, 61
  aqueous, 17, 26, 97, 114
Ammonium chloride, 70
Ammonium molybdate, 171
Ampule, 11
Anisole, 170
Anisaldehyde, 101, 171
Apiezon, 125, 153
Argon (Ar), 22, 31
Asphyxiation, 22
Azeotrope, 69, 70, 79, 80
Azide, 18, 33

Balloon, 35, 76, 83
Barium oxide, 35, 57
Bath
  base or alkali, 153, 175

The Synthetic Organic Chemist's Companion, by Michael C. Pirrung
Copyright © 2007 John Wiley & Sons, Inc.